Stewart Heitmann
Michael Breakspear

Handbook for the
Brain Dynamics Toolbox

Version 2023

Computational Neuroscience

bdtoolbox.org

Brain Dynamics Toolbox
https://bdtoolbox.org

Stewart Heitmann
Victor Chang Cardiac Research Institute
405 Liverpool Street, Darlinghurst NSW 2010, Australia
heitmann@bdtoolbox.org

Michael Breakspear
The University of Newcastle
University Drive, Callaghan NSW 2308, Australia
Michael.Breakspear@newcastle.edu.au

Handbook for the Brain Dynamics Toolbox: Version 2023.
bdtoolbox.org. Sydney, Australia. ISBN 978-0-6450669-3-7.

Preface

Dynamical systems have been fundamental to theoretical neuroscience ever since Alan Hodgkin and Andrew Huxley elucidated the dynamics of spiking neural membranes in the 1950s. Computational neuroscience has progressed enormously since those early days but non-linear differential equations remain at its core. In practice, such equations can only be solved numerically. Numerical methods have thus become an integral part of computational neuroscience. Software toolkits are the manifestation of those endeavours. Each one represents an attempt to balance the shifting tensions between mathematical flexibility and computational convenience. Early toolkits such as GENESIS [2], NEURON [7] and BRIAN [20] focussed on conductance-based models to simulate small networks of neurons with great biological detail. More recently, the Virtual Brain [54] scaled up that approach to the whole brain by combining neural mass models with anatomically derived brain networks (connectomes) to simulate realistic EEG, MEG and fMRI signals. At the other extreme, numerical continuation toolkits such as AUTO [12], XPPAUT [13], MATCONT [11], PyDSToolkit [8] and CoCo [10] provide advanced methods for analysing neuronal dynamics.

Despite this existing corpus of toolboxes, our experience of publishing and teaching computational neuroscience suggests there exists a gap that links theory to practise. In particular, students trained in the varied disciplines of neuroscience often struggle with the practical aspects of translating their interests into computational research. This is particularly true in the fledging fields of computational cognitive and neuroimaging science. The Brain Dynamics Toolbox aims to bridge this gap by allowing those with diverse backgrounds and an interest in neuronal dynamics to explore a variety of models through phase space analysis, time series exploration and other analytic methods. The graphical nature of the toolbox fosters intuitive exploration of the dynamics. The toolbox thus fills the gap between mathematics and biophysics in a deliberately pedagogical manner while retaining the ability to script large-scale simulations and parameter surveys.

This book is for researchers, engineers and students who wish to use the Brain Dynamics Toolbox to construct and explore their own dynamical models. It assumes a basic knowledge of MATLAB programming and some familiarity with the concept of integrating a dynamical system forward in time. Advanced object-oriented programming techniques are not required. The book is complemented by online training courses on the *bdtoolbox.org* website. In particular, the *Toolbox Basics* course which introduces new users to the basic functionality, and the *Modeller's Workshop* which teaches methods for programming bespoke dynamical models.

The software has undergone several major revisions since its initial release in 2016. The graphical interface was overhauled in 2018 to make the display panels more modular. It was rewritten again in 2020 to accommodate MATLAB's new *uitools* interface. This latest version (2023) is a minor update. It removes the dde23a solver which has been superseded by the improved dde23 solver that now ships with Matlab R2023b. The Hilbert panel has also been updated to adjust the mean of the time series to avoid artefacts in the phase angles.

The toolbox has been honoured by the Society for Industrial and Applied Mathematics (SIAM) with a prize in the 2018 Dynamical Systems Software Contest as well as an honourable mention in the 2019 contest. That prize was awarded on the basis of ease of use, effectiveness in solving a problem, genericity of the type of problem the software can tackle, and clarity of documentation — all of which were original design goals of the toolbox. Needless to say, that design benefited greatly from the support and expertise of the members of the Systems Neuroscience Group at QIMR Berghofer Medical Research Institute where the toolbox began. We especially thank Dr Matthew Aburn for his contribution to the design of the SDE solvers.

Australia   *Stewart Heitmann*
November 2023   *Michael Breakspear*

Contents

Chapter 1
Introduction

The *Brain Dynamics Toolbox* is an open-source toolbox for simulating non-linear dynamical systems in MATLAB. It includes a graphical user interface (Figure 1.1) for exploring the dynamics interactively as well as command-line tools for scripting large-scale simulations. The toolbox is intended for researchers and students who wish to use dynamical systems to investigate the theoretical basis of brain function. As such, it supports the three major classes of dynamical systems that typically arise in computational neuroscience — Ordinary Differential Equations (ODEs), Delay Differential Equations (DDEs) and Stochastic Differential Equations (SDEs). Nonetheless the toolbox can be applied to other problem domains too. It makes extensive use of the existing ODE and DDE solvers that are shipped with MATLAB as well as providing several new ones — notably the SDE solvers and the fixed-step Euler method for ODEs. All of which can be applied to Partial Differential Equations (PDEs) using the method lines [55].

The major benefit of the toolbox is flexibility. The graphical interface transforms differential equations into interactive simulations that foster intuitive exploration of the dynamics while still supporting systematic analysis through workspace commands and custom scripts [23]. Moreover, its hub-and-spoke architecture (Figure 1.2) allows unlimited combinations of solvers and plotting panels to be applied to any model with no additional coding effort. This design frees the researcher from the burden of coding bespoke graphical interfaces for each new model that they construct. They can also augment the toolbox with their own custom display panels and solver routines. New modules can be loaded a run-time without having to modify the toolbox source code. The toolbox currently ships with display modules (panels) for visualising mathematical equations, time plots, phase portraits, space-time plots, space-space plots, bifurcation diagrams, Hilbert phases, linear correlations and custom plots for individual models. That list continues to grow with each new version.

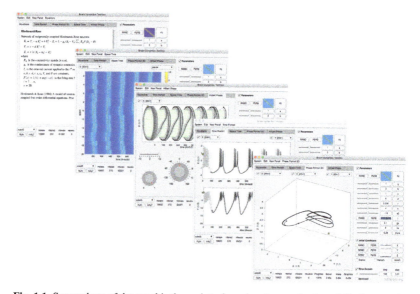

Fig. 1.1 Screenshots of the graphical user interface showing a selection of display panels from the same model — in this case a ring of $n=20$ Hindmarsh-Rose neurons. The model's parameters and initial conditions are controlled via the panel on the right-hand side of the graphical user interface. It is the same in all screenshots. Left to right: (i) Mathematical equations rendered with LaTeX. (ii) Space-time plot of the neural activity on the ring. (iii) Hilbert phases of each neuron's membrane potential. (iv) Time plots of the membrane potential. (v) Phase portrait of the three state variables for a selected neuron.

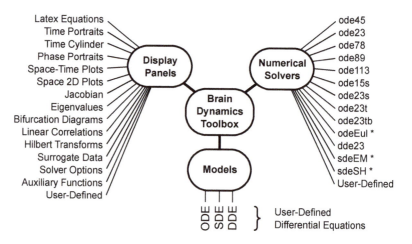

Fig. 1.2 Hub-and-spoke architecture of the Brain Dynamics Toolbox. The toolbox acts as a central hub that connects user-defined models with inter-changeable solver routines and display panels. Solvers marked by an asterisk are unique to the toolbox.

The toolbox uses the same basic approach for solving differential equations as the existing MATLAB solvers (e.g. ode45). The user defines the right-hand side of their dynamical system,

$$\frac{d\mathbf{Y}}{dt} = F(t, \mathbf{Y}, p),$$

as a MATLAB function of the form dYdt=F(t,Y,p) where Y is a vector of state variables and p contains parameter constants. A handle to that function is then passed to the solver which calls it repeatedly to integrate the equations forward in time from a given set of initial conditions.

The Brain Dynamics Toolbox automates the process of calling the solver and plotting the output on the user's behalf. To do so, it requires additional information about the system's state variables and parameters. Those details and others are encapsulated in a specially formatted data structure which we call the *system structure* (Listing 1.1). It is nothing more than a MATLAB structure whose fields follow the conventions of the toolbox. The exact procedures for creating a system structure for a new model are described in Chapters 2-4. In general, that process involves writing a small script which populates the system structure with a handle to the user-defined ODE function (odefun) as well as the names and initial values of the system variables (vardef) and parameters (pardef). The system structure is then loaded into the graphical user interface which runs the model. Various example scripts (Table 1.1) are included in the *bdtoolbox/models* directory. They are all written as conventional MATLAB functions. Advanced object-oriented programming techniques are not required.

Listing 1.1 The system structure for the Hindmarsh-Rose model shown in Figure 1.1.

```
sys =
struct with fields:
      odefun: @odefun
      pardef: [12x1 struct]
      vardef: [3x1 struct]
       tspan: [0 1000]
       tstep: 0.1000
   odesolver: {@ode45  @ode23  @ode113  @odeEul}
   odeoption: [1x1 struct]
      panels: [1x1 struct]
```

Table 1.1 List of example models supplied with the toolbox.

Model	Type	Description
BOLDHRF	ODE	BOLD haemodynamic response function [19].
BrownianMotion	SDE	Geometric Brownian motion.
BTF2003	all	Neural mass with noise and delays [5, 24].
DFCL2009	ODE	Neural mass model of chaotic EEG [9].
EIE0D/1D	ODE	Neural masses with common inhibition [26].
Epileptor2014	O&S	Epileptic seizure dynamics [33].
FisherKolmogorov1D	PDE	Fisher-Kolmogorov equation [15]. †
FitzhughNagumo	ODE	Fitzhugh-Nagumo neuron [16, 44, 53].
FRRB2012	SDE	Multistable neural oscillators with noise [17].
HindmarshRose	ODE	Network of Hindmarsh-Rose neurons [28].
HodgkinHuxley	ODE	Hodgkin-Huxley action potential [30, 22].
HopfieldNet	ODE	Hopfield associative memory network [31].
HopfXY	ODE	Normal form of the Hopf bifurcation.
KloedenPlaten446	SDE	Kloeden & Platen [34] example 4.46.
KuramotoNet	ODE	Network of Kuramoto oscillators [3, 25, 37].
KuramotoSakaguchi	ODE	Kuramoto-Sakaguchi oscillator network [52].
LinearODE	ODE	Linear ODE in two variables.
Lorenz	ODE	The Lorenz attractor [40].
LotkaVolterra	ODE	The Lotka-Volterra predator-prey model.
MorrisLecar	ODE	Morris-Lecar neural membrane [43, 39, 14].
OrnsteinUhlenbeck	SDE	Ornstein-Uhlenbeck noise processes.
Othmer1997	ODE	Intracellular calcium signaling [46, 45].
Pendulum	ODE	Classic damped and driven pendulum.
Pospischil2008	ODE	Minimal Hodgkin-Huxley type neuron [48].
RFB2017	SDE	Neural-mass with multiplicative noise [50].
Strogatz_X_Y_Z	ODE	Examples from Strogatz's textbook [57].
SwiftHohenberg1D	PDE	Swift-Hohenberg Equation [58]. †
Tsodyks1997	ODE	Inhibition-stabilised neural dynamics [60].
VanDerPolOscillators	ODE	Network of Van der Pol oscillators [61, 49].
WaveEquation	PDE	Wave Equations in 1D and 2D. †
WilleBakerEx3	DDE	Example DDE by Willé and Baker [64].
WilsonCowan	ODE	Wilson-Cowan neural masses [65, 66, 29, 27].

† Partial Differential Equations (PDEs) are transformed into Ordinary Differential Equations (ODEs) using the method of lines [55].

1.1 Download and dependencies

The source code for the toolbox is distributed freely under the BSD 2-clause license. It can be downloaded from the Brain Dynamics Toolbox website.

```
https://bdtoolbox.org
```

The toolbox will work with a basic installation of MATLAB R2020a or newer. However, MATLAB R2023b or newer is recommended when solving DDEs due to the improved efficiency of the dde23 solver.

1.2 Installation

Installation is a simple matter of unzipping the source code into a location of your choosing.

```
$ unzip bdtoolbox-2023a.zip
```

The main scripts are located in the top level of the *bdtoolbox* directory. That directory must be in the MATLAB search path. It is also advisable to include the *bdtoolbox/models* directory in the search path if you intend to run any of the example models.

```
>> addpath bdtoolbox-2023a
>> addpath bdtoolbox-2023a/models
```

The display panels and solver routines are installed in the *bdtoolbox/panels* and *bdtoolbox/solvers* directories respectively. The toolbox automatically adds those directories to the search path whenever it is run. User-defined models, panels and solvers need not be located in the installation directory but they must be reachable on the MATLAB search path. Multiple versions of the toolbox can co-exist on the same machine provided that only one version is ever in the search path at any time. The toolbox can be un-installed by simply deleting the *bdtoolbox* directory.

1.3 Getting started

A model is run by loading its system structure into the graphical user interface (bdGUI) . New system structures are typically constructed using a model-specific function for that very purpose. They can also be saved to a *mat* file and loaded from there, as in the following example:

```
>> load HindmarshRose.mat sys
>> bdGUI(sys);
```

In this case, the *HindmarshRose.mat* file is found in the *bdtoolbox/models* directory. The dynamical equations are defined in the *HindmarshRose.m* file in the same directory. The model will fail to load if that file is not reachable on the MATLAB search path.

The bdGUI command will automatically prompt the user to load a model from a *mat* file if it is called with no input parameter. The system structure in that *mat* file is expected to be named sys by convention. Once it loads, the graphical interface automatically computes the solution based on the current parameters in the control panel (Figure 1.3). It also re-computes the solution whenever those controls are adjusted. That behaviour can be temporarily suspended with the HALT button (not shown).

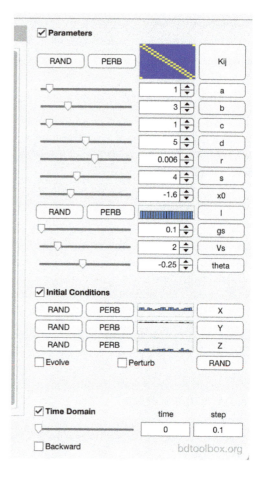

Fig. 1.3 Control panel for the Hindmarsh-Rose model. The model's parameters (Kij, a, b, c, d, r, s, x0, I, gs, Vs, theta) are separated from the initial conditions of the state variables (X, Y, Z). Controls can represent scalars, vectors or 2D matrices. In this case, Kij is a 20×20 matrix. While X, Y, Z and I are each 20×1 vectors. The remaining controls are all scalars. The time domain slider at the bottom of the control panel is used to navigate through time.

The controls themselves can be either scalar, vector or matrix values depending on how the model was defined. Scalar values have edit boxes with slider controls whereas vectors and matrices have RAND and PERB pushbuttons for manipulating their contents. The RAND button assigns uniform random numbers to the individual elements. The PERB button applies a uniform random perturbation to the existing values. Additional editing options can be accessed by clicking the name label of each control. The upper and lower limits of the controls (not shown) are revealed by toggling the checkboxes for each of the *Parameters, Initial Conditions* and *Time Domain*. These limits also dictate the axes plot limits in the various display panels.

Adjusting the time domain

The time slider at the bottom of the control panel (Figure 1.3) is used to navigate the time domain of the simulation. This slider is used by many display panels to segregate the time domain into transient and non-transient parts so that the transient dynamics can be hidden from view. The time span of the slider also defines the time span of the simulation. It is revealed by toggling the *Time Domain* checkbox. The solution is automatically recomputed whenever the time span is changed. The *Backward* checkbox instructs the solver to integrate backwards in time.

Fixed time steps

The computed solution is interpolated onto fixed time steps for the purposes of plotting and analysis. The size of that time step is governed by the step parameter in the control panel (Figure 1.3, bottom right corner). Smaller time steps produce finer plots but at the expense of slower response times. Try to keep the step size as large as practical. The time taken by the interpolator is indicated alongside the solver statistics. If the interpolator is taking longer than the solver then your step parameter is probably too small. For cases where the solver algorithm itself uses fixed time steps (eg. odeEul, sdeEM, sdeSH) then the step parameter also governs the solver step size (*dt*).

Evolving the initial conditions.

The initial conditions can be automatically replaced with the final state of the previous simulation by ticking the *Evolve* checkbox. This is useful for continuing a solution in parameter space or for breaking long simulations into smaller consecutive runs. Use the RUN button to recompute the solution. Use the Bifurcation panel (Section 5.9) to plot an entire branch of solutions.

Perturbing the initial conditions.

The initial conditions can also be automatically perturbed prior to each simulation run by ticking the *Perturb* checkbox. This is useful for exploring the sensitivity of the system to initial conditions or for escaping a solution that is only marginally unstable. The perturbation is uniformly distributed within a range that is equivalent to 5% of the plot limits of each variable. It is applied at the start of each simulation without replacement — meaning that it does not alter the values of the initial conditions in the control panel.

Constructing new system structures.

Some aspects of any model are inevitably fixed at construction. In the previous example, the number of neurons in the Hindmarsh-Rose model was fixed at $n=20$. The *HindmarshRose* function can be used to construct system structures with other configurations.

```
function sys = HindmarshRose(Kij)
```

It takes an $n \times n$ connectivity matrix as input (`Kij`) as input and returns a system structure (`sys`) for a network of n neurons. The following example constructs a system of $n=21$ randomly connected Hindmarsh-Rose neurons and loads it into the graphical user interface.

```
>> n = 21;
>> Kij = rand(n);
>> sys = HindmarshRose(Kij);
>> bdGUI(sys);
```

We can similarly construct a network of $n=47$ neurons using a physiological connectivity matrix from the CoCoMac database [35].

```
>> load cocomac047.mat CIJ
>> sys = HindmarshRose(CIJ);
>> bdGUI(sys);
```

General approach to model configuration.

Most models follow the same approach. The `sys` structure is generated by a model-specific script that determines the size of the model (the number of dynamical equations) from one of the input parameters (usually a connectivity matrix). Use the *help* command for specific details of each model.

```
>> help HindmarshRose
```

1.4 The display panels

Display panels are plug-in modules that can be loaded into the graphical user interface at run-time using the *New Panel* menu. The toolbox ships with a collection of display panels in the *bdtoolbox/panels* directory (Table 1.2). Each one provides a different view of the computed solution.

The *Time Portrait* panel (Figure 1.4) is the principal display panel for inspecting time-series data. It includes two plot axes that operate independently. Each one shows the time course of a selected state variable. If the time series is multi-variate (as is the case here) then the subscripts of the individual time series can be revealed by toggling the checkbox next to the selector. The initial conditions are marked by a yellow hexagon and the end of the trajectory is marked by a filled circle. The transient part of the trajectory — as determined by the position of the time slider in the control panel — is always greyed out. The start of the non-transient time window is marked by an open circle. The transients can be hidden altogether by toggling the *Transients* menu option. The markers can likewise be hidden by toggling the *Markers* menu option.

The vertical plot limits are dictated by the lower and upper limits of the corresponding state variable in the control panel. Those limits can be fitted to the plotted data using the *Calibrate* menu item. The time axes is likewise governed by the limits of the time domain slider. The time series is plotted using fixed time steps unless the *Auto Steps* menu option is ticked, in which case the exact time steps chosen by the solver algorithm are plotted instead. The *Time Points* menu option shows the individual points in the time series. The *Modulo* menu causes the plot lines to wrap around the upper and lower boundaries. The *Hold* menu takes a snapshot of the current plot. The *Clear* menu erases all snapshots. The *Undock* menu moves the panel into a separate window. The *Close* menu closes the display panel. These same conventions are followed by all display panels, where appropriate.

Table 1.2 The display panel classes.

Panel Class	Menu Name	Description
bdAuxiliary	Auxiliary	Model-specific plotting functions.
bdBifurcation	Bifurcation 2D	Bifurcation diagram in one state variable.
bdBifurcation3D	Bifurcation 3D	Bifurcation diagram in two state variables.
bdCorrPanel	Correlation	Linear correlation for multi-variate data.
bdDFDY	dFdY	The Jacobian matrix of the ODE.
bdEigenvalues	Eigenvalues	Eigenvalues of the Jacobian matrix.
bdHilbert	Hilbert	Hilbert phase of a selected state variable.
bdLatexPanel	Equations	Mathematical equations using LaTeX.
bdPhasePortrait	Phase Portrait 2D	Phase portrait in two state variables.
bdPhasePortrait3D	Phase Portrait 3D	Phase portrait in three state variables.
bdSolverPanel	Solver	Solver step-size and error tolerances.
bdSpace2D	Space 2D	Space plots for 2D state variables.
bdSpaceTime	Space-Time	Space-time plots for multi-variate data.
bdSurrogate	Surrogate	Phase randomised surrogate data.
bdSystemLog	System Log	Debug log of system events.
bdTimeCylinder	Time Cylinder	Time plot for polar variables.
bdTimePortrait	Time Portrait	Time plots for two selected state variables.

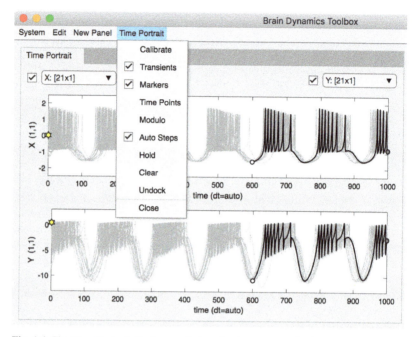

Fig. 1.4 The Time Portrait display panel plots the time course of selected state variables. In this case the selected variables ('X' and 'Y') are both multi-variate state variables from a network of $n = 21$ Hindmarsh-Rose neurons.

1.5 The workspace interface

The bdGUI command returns a handle to the internal states of the graphical user interface that it creates.

```
>> gui = bdGUI(sys);
```

The public properties of that handle (Table 1.3) provide a direct interface between the user's private workspace and the running simulation. Listing 1.2 shows how to use the handle to read the state of the running model and to plot the computed solution in a new figure (Figure 1.5). The data that is returned from the gui handle belongs exclusively to the user's workspace and can be passed to any third-party application in the same manner.

Nonetheless it is worth emphasising that the gui object is a *handle* to the internal states of the graphical user interface rather than a separate copy. So any changes to its properties are instantly reflected in the graphical user interface and vice versa. The parameters of the running model can thus be modified via gui.par. Likewise, the initial conditions can be modified via gui.var0. The computed trajectories are accessible (read-only) from gui.vars and their time points are in gui.t.

Listing 1.2 Using the public properties of the bdGUI class to access the current state of parameters and variables in the Hindmarsh-Rose model. The output of the plot command (line 27) is shown in Figure 1.5.

```
>> load HindmarshRose.mat sys
>> gui = bdGUI(sys);
>> gui.par
ans =
   struct with fields:
        Kij: [20x20 double]
          a: 1
          b: 3
          c: 1
          d: 5
          r: 0.0060
          s: 4
         x0: -1.6000
          I: [20x1 double]
         gs: 0.1000
         Vs: 2
      theta: -0.2500

>> gui.vars
ans =
   struct with fields:
      X: [20x1001 double]
      Y: [20x1001 double]
      Z: [20x1001 double]

>> figure;
>> plot(gui.t,gui.vars.X);
```

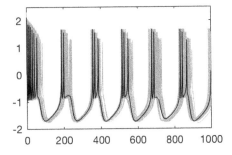

Fig. 1.5 The output produced by Listing 1.2.

Table 1.3 Public properties of the bdGUI class.

Property	Type	Description	Access
version	string	Toolbox version string '2023a'.	read-only
par	struct	Parameters of the model.	read-write
lag	struct	DDE lag parameters.	read-write
var0	struct	Initial conditions.	read-write
vars	struct	Computed trajectories.	read-only
tspan	double	Time span of the simulation.	read-write
tstep	double	Interpolation time step.	read-write
tval	double	Value of the time slider.	read-write
t	double	Time steps returned by the solver.	read-only
sys	struct	Current state of the model's sys structure.	read-only
sol	struct	Current solution as returned by the solver.	read-only
panels	struct	Properties of the individual display panels.	read-only
halt	logical	State of the HALT button.	read-write
evolve	logical	State of the EVOLVE button.	read-write
perturb	logical	State of the PERTURB button.	read-write
fig	figure	Graphics handle to the application figure.	read-write

Workspace scripting.

The workspace interface is particularly convenient for scripting quick parameter surveys. The following brief script illustrates how to ramp the current (I) applied to all neurons in the Hindmarsh-Rose model. In this case the gui.par.I parameter is a vector of n values (one per neuron) which together are ramped from 0 to 2 in increments of 0.1.

```
n = 21;
Kij = rand(n);
sys = HindmarshRose(Kij);
gui = bdGUI(sys);
for I=0:0.1:2
    gui.par.I = I*ones(n,1);
end
```

The graphical interface automatically recomputes the solution each time the gui.par.I parameter is assigned a new value. MATLAB will wait for that computation to complete before performing the next instruction in the script. Hence there is no need to insert a *wait* command in the for-loop.

1.6 The command-line tools

The toolbox includes a suite command-line tools (Chapter 7) for scripting large-scale simulations without having to invoke the graphical user interface. These tools provide the same level of control as the workspace interface but are intended for large parameter surveys that run in batch mode. The primary command-line tool is `bdSolve` which solves an initial-value problem for a given system structure. The `bdEval` command interpolates the computed solution for a given set of time points. It is analogous to the MATLAB `deval` command except that it also works for solutions returned by the toolbox's custom solver routines. The toolbox also has some useful utility functions such as `bdGetValue` and `bdSetValue` which allow the parameters and initial conditions in the `sys` structure to be accessed by name. Another important tool is `bdSysCheck` which validates the format of a user-defined system structure during the development cycle.

1.7 A worked example

We complete this introductory chapter by simulating traveling waves in a ring of Hindmarsh-Rose [28] neurons using the example model shipped with the toolbox. It is a model of spike-bursting behaviour where each neuron is represented by its membrane potential $X_i(t)$, a fast recovery variable $Y_i(t)$ and a slow recovery variable $Z_i(t)$.

$$\dot{X}_i = Y_i - aX_i^3 + bX_i^2 - Z_i + I_i - g_s(X_i - V_s)\sum K_{ij}F(X_j - \theta)$$
$$\dot{Y}_i = c - dX_i^2 - Y_i$$
$$\dot{Z}_i = r(s(X_i - x_0) - Z_i)$$

The two recovery variables represent ionic currents in the neural membrane. The fast current contributes to the shape of each spike and the slow current contributes to the envelope of the spike bursting. The connectivity matrix K_{ij} defines the topology of the network and the synaptic conductance g_s governs the strength of the coupling between neurons. The parameter I_i represents an external stimulation current. It is applied to each neuron independently. The other parameters are not important here, so we omit them for brevity.

We begin by constructing an instance of the Hindmarsh-Rose model using a connectivity matrix (`Kij`) that defines a ring network.

```
>> n = 21;
>> Kij = circshift(eye(n),1) ...
         + circshift(eye(n),-1);
>> sys = HindmarshRose(Kij);
```

The `sys` structure is loaded into the graphical interface in the usual manner.

```
>> gui = bdGUI(sys);
```

We wish to drive only one neuron in the ring and observe its influence on its neighbouring neurons. For demonstration purposes, we use the `gui` handle rather than the graphical control panel to apply a non-zero current to the 11th neuron in the ring.

```
>> gui.par.I(11) = 2;
```

That level of current induces repetitive bursts of spikes in the membrane potential of the stimulated neuron (X_{11} in Figure 1.6a) but those spikes fail to propagate to the neighbouring neurons in the ring (as shown by the space-time panel in Figure 1.6b) because the synaptic conductance is too weak ($g_s=0.1$). Using the slider control, we gradually increase the synaptic conductance and observe the emergence of spreading activity (Figure 1.6c) at $g_s=0.37$. That spreading activity attenuates rapidly with distance. It is not until the synaptic conductance is increased to $g_s=0.4$ that the model supports self-sustaining propagating waves (Figure 1.6d). They travel around the ring in opposite directions until they collide and annihilate one another. Even then those waves are only emitted from the stimulation site on every second burst cycle. Increasing the synaptic conductance of the network further still, we observe waves emitted every 2:3 cycles for $g_s=0.41$ (Figure 1.6e) and on every cycle for $g_s=0.42$ (Figure 1.6f). Unwanted transients can be eliminated from successive simulations by using the *Evolve* button to gradually follow the solution from $g_s=0.1$ to $g_s=0.42$.

This brief exercise illustrates how the toolbox can be used to quickly simulate and explore complex dynamical phenomenon in an existing model. The remaining chapters describe how to use the toolbox to build and run user-defined dynamical models.

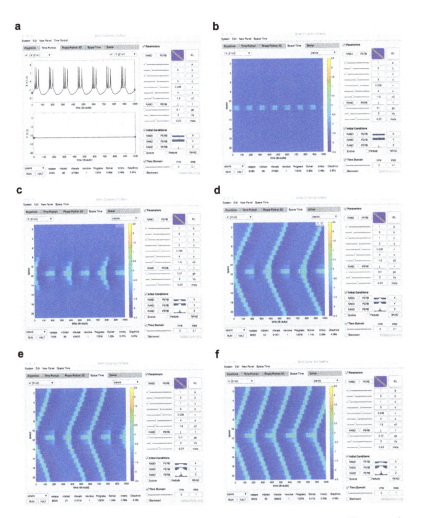

Fig. 1.6 Propagating waves in a ring of $n=21$ Hindmarsh-Rose neurons. (**a**) Time portraits for the stimulated neuron (X_{11}) in the upper panel and its neighbouring neuron (X_{12}) in the lower panel. The synaptic conductance is $g_s=0.1$. (**b**) Space-time plot of $X_i(t)$ for all neurons in the ring when $g_s=0.1$. The horizontal axis is time. The vertical axis represents the spatial position of each neuron and has periodic boundary conditions. The colour scale indicates the amplitude of $X(t)$. (**c-f**) Space-time plots for the same model with $g_s=0.37$, $g_s=0.40$, $g_s=0.41$ and $g_s=0.42$ respectively.

Chapter 2
Ordinary Differential Equations

Ordinary Differential Equations (ODEs) have the general form,

$$\frac{dY}{dt} = F(t, Y)$$

where $F(t, Y)$ is a function of the dynamic variable $Y(t) \in \mathbb{R}^n$ and time t. ODEs are typically posed as initial value problems — wherein the initial state Y_0 is known at some time $t=t_0$ and the evolution of $Y(t)$ is sought for some time span $t_0 \leq t \leq t_1$ by the process of forward integration. Such problems are solved numerically in MATLAB using the family of ODE solvers ode45, ode23s, ode113, ode15s, etc. The individual solvers in this family are optimized for different types of problems but they all operate in the same manner. Specifically, they require the user to implement their ODE as a MATLAB function of the form

```
function dYdt = F(t,Y,a,b,c,...)
```

where Y is the state variable (vector) and a, b, c, ... are model-specific parameters. The ODE solver, say ode45, takes a handle to that function which it calls at each step of the forward integration.

```
tspan = [0 1];              % time span [t0 t1]
Y0 = rand(n,1);             % initial conditions
options = odeset('RelTol',1e-6);
sol = ode45(@F,tspan,Y0,options,a,b,c,...);
```

The solver returns the computed solution (sol.y) and the corresponding time points (sol.x) as a structure. The time steps taken by the solver may be irregular, so it is common to interpolate the solution onto a regularly spaced grid using the deval function.

```
t = 0:0.01:1;               % time domain
Y = deval(sol,t);           % interpolate
```

17

The Brain Dynamics Toolbox uses those same MATLAB ODE solvers (as well as some new ones) but it automates the process of calling the solver and plotting the output. The user must still implement the ODE function as before. Only now that function handle is passed to the toolbox instead of to the ODE solver.

2.1 Defining an ODE

We demonstrate by implementing the linear system of ODEs,

$$
\begin{aligned}
\dot{x} &= ax + by, \\
\dot{y} &= cx + dy,
\end{aligned}
\tag{2.1}
$$

where $x(t)$ and $y(t)$ are the dynamic variables and a, b, c, d are constants. The dot notation indicates time derivatives. The simplicity of these equations allows us to focus on the implementation details rather than the mathematics.

The source code for implementing these equations (2.1) with the toolbox is given in Listing 2.1. It corresponds to the *LinearODE* example model (Figure 2.1). The script works by constructing a system structure (`sys`) that it populates with specific fields (lines 3–33). It is run as follows.

```
>> sys = LinearODE();
>> gui = bdGUI(sys);
```

At a minimum, the `sys` structure for an ODE must contain the fields `odefun`, `pardef`, `vardef` and `panels` (Table 2.1). The `odefun` field defines a handle to the user-defined `F(t,Y,a,b,c,d)` function (lines 36–41). The `pardef` and `vardef` fields define the names and initial values for the model's parameters and variables. Finally, the `panels` field (lines 16–33) defines which display panels to load at start-up.

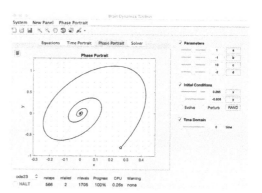

Fig. 2.1 Screenshot of bdGUI running the LinearODE model.

Listing 2.1 Definition for the linear system of ODEs given by equations (2.1).

```matlab
% The system definition function
function sys = LinearODE()
    % Handle to our ODE function
    sys.odefun = @odefun;

    % ODE parameter definitions
    sys.pardef = [ struct('name','a', 'value', 1);
                   struct('name','b', 'value',-1);
                   struct('name','c', 'value',10);
                   struct('name','d', 'value',-2) ];

    % ODE variable definitions
    sys.vardef = [ struct('name','x', 'value', 0.5);
                   struct('name','y', 'value',-0.1) ];

    % Latex (Equations) panel
    sys.panels.bdLatexPanel.title = 'Equations';
    sys.panels.bdLatexPanel.latex = {
        '\textbf{LinearODE}';
        '';
        'System of linear ordinary differential equations';
        '  $\dot x(t) = a\,x(t) + b\,y(t)$';
        '  $\dot y(t) = c\,x(t) + d\,y(t)$';
        'where $a, b, c, d$ are scalar constants.' };

    % Time Portrait
    sys.panels.bdTimePortrait = [];

    % Phase Portrait
    sys.panels.bdPhasePortrait = [];

    % Solver Panel
    sys.panels.bdSolverPanel = [];
end

% The ODE function.
function dYdt = odefun(t,Y,a,b,c,d)
    % Y and dYdt are both (2x1) vectors.
    % Parameters a,b,c,d are scalars.
    dYdt = [a b; c d] * Y;              % matrix multiplication
end
```

Table 2.1 Mandatory fields for an ODE `sys` structure.

Field	Type	Description
odefun	@F(t,Y,a,b,c,...)	Handle to the user-defined ODE function
pardef	struct array	Defines the ODE parameters: a,b,c,...
vardef	struct array	Defines the ODE variables: x,y
panels	struct	Configuration options for the display panels

2.2 The ODE function

The ODE function defines the system of equations that are to be solved. It follows the same format as that required by the MATLAB ODE solvers (ode45, ode23, etc). Specifically, it must have the syntax

```
function dYdt = odefun(t,Y,a,b,c,...)
```

where t is a scalar, Y and dYdt are both n×1 vectors. The parameters a,b,c,... may be scalars, vectors or matrices. In our example, the ODE function (lines 36–41 of Listing 2.1) is defined as,

```
function dYdt = odefun(t,Y,a,b,c,d)
    dYdt = [a b; c d] * Y;
end
```

where the parameters a, b, c, d are all scalar values and parameter t is unused. The ODE solver calls this function at each time step of the forward integration. It passes the state variables $x(t)$ and $y(t)$ as the column vector Y=[x(t);y(t)] and expects the function to return the time derivatives as the corresponding column vector dYdt=[ẋ(t);ẏ(t)]. The equations (2.1),

$$\begin{bmatrix} \dot{x}(t) \\ \dot{y}(t) \end{bmatrix} = \begin{bmatrix} a & b \\ c & d \end{bmatrix} \begin{bmatrix} x(t) \\ y(t) \end{bmatrix},$$

are implemented using matrix multiplication (line 40 of Listing 2.1).

The toolbox is given a handle to the ODE function via the odefun field of the sys structure (line 4 of Listing 2.1).

```
sys.odefun = @odefun;
```

The toolbox uses the odefun handle to call the relevant ODE solver on the user's behalf.

ODE functions can be difficult to debug. The `bdSysCheck` function makes the task easier by validating the ODE function parameters against the definitions in the `sys` structure. It also exercises the ODE function by passing it to the relevant ODE solvers. See Chapter 7.2 for more about system validation.

2.3 The ODE parameters

The names and values of the ODE parameters are defined in a specially formatted array of structures which is called `pardef` (lines 6–10 of Listing 2.1).

```
% ODE parameter definitions
sys.pardef = [struct('name','a',  'value', 1);
              struct('name','b',  'value',-1);
              struct('name','c',  'value',10);
              struct('name','d',  'value',-2)];
```

The array is initialized with one element for each user-defined parameter expected by the ODE function (See Listing 2.2 for alternative syntaxes). The `name` field of the `pardef` structure defines the name of the parameter as it appears in the graphical controls. It is also used by the command line tools to access parameter values by name (Chapter 7). The `value` field defines the parameter's default value. In this case, there are four scalar parameters with default values a=1, b=-1, c=10 and d=-2. In general, the `value` field may be assigned a scalar, vector or matrix value provided that its size matches that expected by the ODE function (line 37 of Listing 2.1).

The `name` and `value` fields are the two most important fields of the `pardef` structure. Indeed they are both mandatory. The `pardef` structure also accepts an optional field called `lim`. It contains two values `[lo hi]` which define the lower and upper limits of the parameter as used by the plot axes and the slider control. It is these values that are adjusted by the *Calibrate Axes* menu item. In most cases the toolbox uses sensible default values when the `lim` field is left undefined, so it can usually be omitted from the system definition.

2.4 The ODE variables

The state variables of the ODE are defined in the `vardef` array of structs (lines 12-14 of Listing 2.1) using the same conventions as the `pardef` array.

```
% ODE variable definitions
sys.vardef = [struct('name','x',  'value', 0.5);
              struct('name','y',  'value',-0.1)];
```

In this case there are two state variables (x=0.5, y=-0.1) whose values represent the initial conditions of the ODE. The solver passes these values to the ODE function as a column vector (Y0=[0.5;-0.1]). In cases where the `vardef` array contains a mix of scalar, vector and matrix values then those values are recombined into one monolithic vector (Y0). The `bdGetValues` function can be used to extract the initial values from a `vardef` structure in a form suitable for passing to an ODE solver.

The `vardef` structure includes an additional field called `solindx` which contains the indices of the named variable in the `sol` structure that the solver returns. This field is used by the display panels to map the name of an ODE variable to the appropriate rows of the solution data. The `solindx` field is automatically populated by the toolbox when the model is loaded. So it should not be explicitly defined by the user.

Listing 2.2 Three equivalent coding styles for initializing an array of `pardef` structs. We prefer method 3 because it is compact and avoids indexing errors.

```
 1  % method 1
 2  sys.pardef(1).name='a';   sys.pardef(1).value= 1;
 3  sys.pardef(2).name='b';   sys.pardef(2).value=-1;
 4  sys.pardef(3).name='c';   sys.pardef(3).value=10;
 5  sys.pardef(4).name='d';   sys.pardef(4).value=-2;
 6
 7  % method 2
 8  sys.pardef(1) = struct('name','a', 'value', 1);
 9  sys.pardef(2) = struct('name','b', 'value',-1);
10  sys.pardef(3) = struct('name','c', 'value',10);
11  sys.pardef(4) = struct('name','d', 'value',-2);
12
13  % method 3
14  sys.pardef = [ struct('name','a', 'value', 1);
15                 struct('name','b', 'value',-1);
16                 struct('name','c', 'value',10);
17                 struct('name','d', 'value',-2) ];
```

2.5 Display Panel options

It is often convenient for a model to automatically load some well-chosen panels at start-up. For this reason bdGUI automatically loads those display panels which are explicitly named in the panels field of the model's system structure. In most cases, it is enough to define the name of the panel class as an empty field.

```
sys.panels.bdTimePortrait = [];
```

A list of those names is provided in Table 1.2. Additional display panel options can also be specified within the class name if required.

```
sys.panels.bdTimePortrait.title='Time Portrait';
sys.panels.bdTimePortrait.grid=true;
```

The meaning of those options are specific to each display panel (See Chapter 5). The default options for each panel are usually adequate so there is rarely any reason to define them in the system structure.

Our *LinearODE* example (Listing 2.1) nominates the bdLatexPanel, bdTimePortrait, bdPhasePortrait and bdSolverPanel panels for automatic loading in lines 16–33. All apart from the bdLatexPanel panel are specified as empty fields.

Configuring the LaTeX Equations Panel

The latex option for the bdLatexPanel panel is a common example of a panel option that must be explicitly defined in the system structure. The bdLatexPanel renders mathematical equations in the graphical user interface using the built-in LaTeX interpreter. The latex option defines the input to the LaTeX interpreter as a cell array of strings. In our example ODE (lines 16–24 of Listing 2.1)

```
sys.panels.bdLatexPanel.latex = {
    '\textbf{LinearODE}';
    '';
    'System of linear ordinary differential
    equations';
    '   $\dot x(t) = a\,x(t) + b\,y(t)$';
    '   $\dot y(t) = c\,x(t) + d\,y(t)$';
    'where $a, b, c, d$ are scalar constants.'};
```

the LaTeX strings produce the typeset equations shown in Figure 2.2. Be aware that bdLatexPanel expects those strings to be formatted for the MATLAB built-in *LaTeX* interpreter rather than the *TeX* interpreter. Refer to the MATLAB LaTeX documentation for the syntax of specific commands.

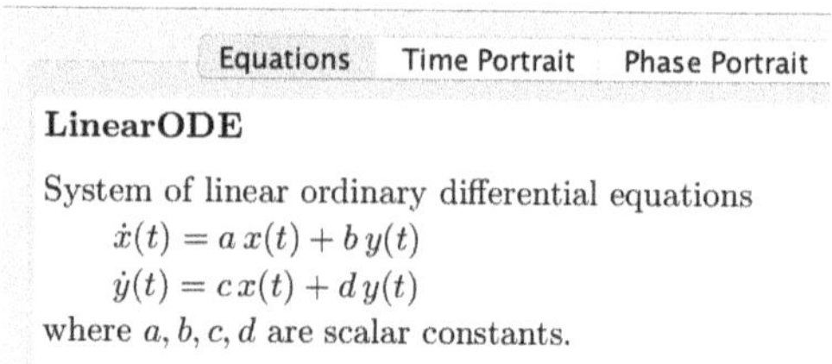

| Equations | Time Portrait | Phase Portrait |

LinearODE

System of linear ordinary differential equations

$$\dot{x}(t) = a\,x(t) + b\,y(t)$$
$$\dot{y}(t) = c\,x(t) + d\,y(t)$$

where a, b, c, d are scalar constants.

Fig. 2.2 Detail of the La-TeX output produced by the `bdLatexPanel` in the `LinearODE` model.

2.6 ODE solvers and options

The toolbox supports multiple ODE solver routines (Table 2.2) which can be applied interchangeably at run-time via a pull-down menu. The `odesolver` field can be used to customize that list by adding or removing handles to the solver routines. Handles to user-defined solver routines can also be included.

```
sys.odesolver = {@ode45, @ode23s, @odeEul};
```

Different solver routines are optimized for different problem domains and finding the best one is usually a matter of trial and error. The `ode45` solver is recommended for most problems. Whereas stiff problems — where the state variables change rapidly — may do better with `ode15s` or any of the `ode23s`, `ode23t`, `ode23tb` family of solvers. See *'Choose an ODE solver'* in the MATLAB documentation for more details.

The `odeEul` solver is specific to the Brain Dynamics Toolbox. It implements the fixed-step Euler method which is inefficient but remains popular in the literature because it is simple to implement.

Table 2.2 ODE solver routines.

Solver	Description
ode45	General-purpose solver with medium accuracy for non-stiff solutions.
ode23	Suitable for moderately stiff solutions with low accuracy requirements.
ode78	Suitable for smooth solutions with high accuracy requirements. †
ode89	Suitable for very smooth solutions with high accuracy requirements. †
ode113	For problems that are expensive to compute or have strict error tolerances.
ode15s	Suitable for stiff solutions with low or medium accuracy requirements.
ode23s	Suitable for stiff solutions with low accuracy and a known Jacobian.
ode23t	Suitable for stiff solutions that benefit from a lack of numerical damping.
ode23tb	Alternative to ode23s that may be more efficient in some circumstances.
odeEul	Fixed-step Euler method. Inefficient but popular due to its simplicity.

† Introduced in MATLAB R2021b.

Solver options

The efficiency of most solver routines can be dramatically improved with carefully chosen solver options. MATLAB provides the odeset and odeget functions for setting and getting ODE solver options. The Brain Dynamics Toolbox uses those same solver options but they are instead passed to the solver via the odeoption field of the sys structure. Its contents can be manipulated using the conventional odeset and odeget functions. However it is often simpler to directly assign an option to the structure using its field name, as in the following example.

```
% example ODE options
sys.odeoption.AbsTol = 1e-6;
sys.odeoption.RelTol = 1e-3;
```

The full list of ODE solver options is given in Section 2.8. The odeEul solver only recognizes the InitialStep, OutputFcn and OutputSel options. The InitialStep option is mandatory for the odeEul solver as it defines the step size of the method.

2.7 Specifying a Jacobian

Most ODE solvers do not require an explicit form of the Jacobian but its inclusion can improve performance for large systems of equations. It is especially beneficial for ode23s. The toolbox requires the Jacobian to be defined as a user-defined function. A handle to that function is passed to the ODE solver via the Jacobian field of the odeoption structure.

```
sys.odeoption.Jacobian = @jacfun;
```

That function accepts the same input parameters as odefun and returns the Jacobian matrix as its output. In the case of our linear system of equations (2.1) the Jacobian matrix is

$$\begin{bmatrix} \partial f_x & \partial f_y \\ \partial g_x & \partial g_y \end{bmatrix} = \begin{bmatrix} a & b \\ c & d \end{bmatrix}$$

and the corresponding Jacobian function is

```
function J = jacfun(t,Y,a,b,c,d)
    J = [a b; c d];
end
```

The Jacobian function has been omitted from Listing 2.1 for brevity but it is included in the LinearODE example code that is shipped with the toolbox in the models directory.

2.8 System structure for ODEs

This section provides a complete list of all `sys` fields that are relevant to ODEs. See Chapter 5 for fields related to display panels.

`sys.odefun = @(t,Y,a,b,c,...)`

A handle to a user-defined function of the form

`dYdt = myfun(t,Y,a,b,c,...)`

where `t` is a scalar and both `dYdt` and `Y` are n×1 vectors. The user-defined parameters `a, b, c, ...` may be scalars, vectors or matrices. The function defines the right-hand side of the ODE. It has the same syntax required by the MATLAB ODE solvers (e.g. `ode45`).

`sys.pardef`

An array of structures that define the names and values of each parameter in the user-defined ODE. There must be one array element for each user-defined parameter (`a,b,c,...`) in `odefun`. The structure has the following fields:

`name = <string>`

The name of the ODE parameter as it appears in the graphical interface. It is also used by the `bdGetValue` and `bdSetValue` functions to read and write parameter values by name (Chapter 7).

`value = <numeric>`

The default value of the ODE parameter. It may be either a scalar, vector or matrix provided that its size matches that of the corresponding input parameter of `odefun`.

`lim = [lo hi]`

The lower and upper limits of the ODE parameter (optional). These limits are applied to the plot axes in the display panels as well as the sliders in the control panel. If no limits are specified then they are inferred from the default value(s) of the parameter.

`sys.vardef`

An array of structures that define the names and values of each ODE state variable. Each structure has the following fields:

`name = <string>`

The name of the ODE state variable as it appears in the graphical interface. The name is also used by the `bdGetValue` and `bdSetValue` functions (Chapter 7) which can be used in scripts to read and write initial values by name.

```
value = <numeric>
```
 The initial value of the ODE state variable. The numeric value may
 be formatted as either a scalar, vector or matrix.

```
lim = [lo hi]
```
 The lower and upper limits of the state variable (optional). These
 limits are applied to the plot axes in the display panels as well as
 the sliders in the control panel. If no limits are specified then they
 are estimated from the initial values of the state variable.

```
solindx = <integer>
```
 An array of indices that map each element of the state variable to
 the matching rows of the solution data (`sol.y`) produced by the
 solver. This field is initialised by bdGUI whenever it loads a system
 structure so there is no need to define it explicitly. It is used by
 display panels to extract the relevant solution data for plotting.

sys.panels

 A structure containing the various display panel options. The fields within
 this structure correspond to the names of the display panel classes. The
 options for each display panel are described in Chapter 5.

sys.tspan = [t0 t1]

 The time span of the numerical integration where $t0 \leq t1$ (optional).

sys.tval = t0

 The value of the time slider where $t0 \leq tval \leq t1$ (optional).

sys.tstep = 1

 The time step used by the interpolator (optional).

sys.odesolver = {@ode45, @ode23, @ode113, @ode15s, @ode23s, @ode23t, @ode23tb, @odeEul}

 A cell array of function handles to the relevant ODE solvers (optional).

sys.odeoption

 A structure containing the ODE solver options (optional). It uses the same
 format as the MATLAB `odeset` and `odeget` functions. Those options
 and their default values are summarised here. See the `odeset` documentation for a complete description.

```
RelTol = 1e-3
```
  Error tolerance relative to the magnitude of each solution compo-

nent. It controls the correct number of digits in the solution except where the solution is smaller than `AbsTol`.

`AbsTol = 1e-6`

Absolute tolerance of the solver error. The precision of solution components below this tolerance threshold is not considered to be important.

`NormControl = 'off'`

Use the norm of the solution rather than the absolute value of each component when computing the relative tolerance of the solver error.

`NonNegative`

Specifies those solution components that are always non-negative. Not available for `ode23s` or `ode15i`. Not available for `ode15s`, `ode23t`, `ode23tb` when a mass matrix is used.

`OutputFcn`

Handle to a user-supplied function that is called at each time step. Available with `bdSolve` but not `bdGUI`.

`OutputSel`

Specifies which solution components are passed to `OutputFcn`. Available with `bdSolve` but not `bdGUI`.

`Refine`

Not used by either `bdSolve` or `bdGUI`.

`Stats`

Solver statistics. Available with `bdSolve` but not `bdGUI`.

`InitialStep`

Sets an upper bound on the initial step taken by the solver. This field is mandatory for the `odeEul` solver where it defines the size of every step taken by the solver.

`MaxStep`

Sets an upper bound on each step taken by the solver.

`Events`

Handle to a user-supplied function that detects points of interest in the solution.

`Jacobian`

A matrix or function that returns the Jacobian of the ODE. It is crucial to the performance of stiff ODE solvers (`ode15s`, `ode23s`, `ode23t`, `ode23tb` and `ode15i`).

`JPattern`

Sparsity pattern of the Jacobian matrix.

`Vectorized=off`
> Indicates that the `sys.odefun` function is vectorized.

`Mass`
> Mass matrix for problems of the form $My' = f(t,y)$.

`MStateDependence`
> State dependence of the mass matrix.

`MvPattern`
> Sparsity pattern of the mass matrix.

`MassSingular`
> Specifies whether the mass matrix should be tested for singularity.

`InitialSlope`
> Slope of the solution components at the initial conditions.

`sys.halt = false`
`sys.evolve = false`
`sys.perturb = false`
> The initial states of the HALT, EVOLVE, PERTURB buttons.

`sys.UserData`
> This field is reserved for user-defined data. It can be used to store model-specific function handles and data structures directly in the system structure. The contents of this field are not modified by the toolbox and there are no restrictions on what can be stored in it.

`sys.auxfun = @(t,Y,a,b,c,...)`
> This field was deprecated in version 2018a.

`sys.self = @()`
> This field was deprecated in version 2018a.

Chapter 3
Delay Differential Equations

Delay Differential Equations (DDEs) with constant delays have the general form,

$$\frac{dY(t)}{dt} = F\left(t, Y(t), Y(t-\tau_1), Y(t-\tau_2), \ldots, Y(t-\tau_k)\right)$$

where F is a function of the dynamic variable $Y(t) \in \mathbb{R}^n$ at time t and also at selected time delays $\tau_1, \tau_2, \ldots, \tau_k$. Such DDEs can be solved numerically using the MATLAB dde23 solver. It follows the same approach as the ODE solvers (ode45, ode23, etc) except that F must be defined with the syntax

```
function dYdt = F(t,Y,Z,a,b,c,...)
```

where the vector Y contains the current value of $Y(t)$ and the columns of the matrix Z contain consecutive time-delayed values $Y(t-\tau_i)$. As always, the parameters a, b, c, . . . are model-specific.

Like the ODE solver, the DDE solver calls the user-defined F function at each step of the forward integration. The solver must also be initialised with the historical values of $Y(t)$ prior to the initial time point ($t = t_0$). The dde23 solver supports three methods of doing this. The first method assumes that $Y(t)$ is constant for $t < t_0$, The second method requires the historical values of $Y(t)$ to be supplied as a matrix of column vectors. The third method requires a user-defined function that returns the historical value of $Y(t)$ for any given time. Of these three methods, the Brain Dynamics Toolbox currently only supports constant historical values.

The basic method used by the toolbox to call the DDE solver is illustrated in Listing 3.1. It requires a handle to the DDE function (@F), as well as the model parameters (a, b, c, d), time lags ($\tau_1=1, \tau_2=0.5, \tau_3=0.2$) and historical values for $Y(t)$. All of which must be defined in the model's system structure. The method for constructing the system definition for a DDE is similar to that for an ODE (Chapter 2) but with some distinctions that are illustrated next.

31

Listing 3.1 Solving a DDE with constant historical values using the MATLAB dde23 solver. In this hypothetical example, the function handle @F refers to a user-defined DDE that defines the dynamical behaviour of $Y(t) \in \mathbb{R}^n$. It has three time lags ($\tau_1=1$, $\tau_2=0.5$, $\tau_3=0.2$) and four model-specific parameters (a, b, c, d) whose definitions are not shown. The simulation time spans $0 \le t \le 1$. The history vector defines the unvarying values of $Y(t)$ for all $t < 0$. Those values are chosen at random in this example.

```
1   lags = [1 0.5 0.2];                      % time lags
2   history = rand(n,1);                     % initial conditions
3   tspan = [0 1];                           % time span [t0 t1]
4   options = ddeset('RelTol',1e-6);         % DDE solver options
5   sol = dde23(@F,lags,history,tspan,options,a,b,c,d);
```

3.1 Defining a DDE

We illustrate the method using the delay equations from example 3 of Willé and Baker [64]. Specifically,

$$a \dot{y}_1(t) = y_1(t-1), \tag{3.1}$$

$$b \dot{y}_2(t) = y_1(t-1) + y_2(t-0.2), \tag{3.2}$$

$$c \dot{y}_3(t) = y_2(t), \tag{3.3}$$

where the dynamic variables $y(t)$ are subject to time lags $\tau_1=1$ and $\tau_2=0.2$. It corresponds to the ddex1 example code that is shipped with MATLAB as well as the *WilleBakerEx3* model that is shipped with the Brain Dynamics Toolbox (Table 1.1). The time scale constants ($a=1$, $b=1$, $c=1$) are not present in the original model [64] but have been included here to illustrate the use of additional parameters. The state variables are assumed to have constant values $y_1=y_2=y_3=1$ for $t<0$.

Listing 3.2 shows a minimal system definition for this model and a screen-shot of the graphical interface is shown in Figure 3.1. The system definition returns a sys structure in which the handle to the user-defined DDE function (lines 41–49) is returned in the field ddefun. The model parameters (pardef)) and state variables (vardef) are defined using the same conventions as ODEs (Chapter 2). The lag parameters (lagdef) also follow those same conventions, albeit lags are specific to DDEs. The remainder of Listing 3.2 (lines 19–37) concerns the display panel settings which are the same for DDEs as for ODEs (Chapter 2.5). Individual display panels are described in detail in Chapter 5.

Listing 3.2 System definition for the delay equations (3.1–3.3) of Willé and Baker [64].

```matlab
function sys = WilleBakerEx3()
    % Handle to our DDE function
    sys.ddefun = @ddefun;

    % DDE parameters
    sys.pardef = [ struct('name','a', 'value', 1)
                   struct('name','b', 'value', 1)
                   struct('name','c', 'value', 1) ];

    % DDE lag parameters
    sys.lagdef = [ struct('name','tau1', 'value',1.0)
                   struct('name','tau2', 'value',0.2) ];

    % DDE state variables
    sys.vardef = [ struct('name','y1', 'value',1)
                   struct('name','y2', 'value',1)
                   struct('name','y3', 'value',1) ];

    % Latex (Equations) panel
    sys.panels.bdLatexPanel.title = 'Equations';
    sys.panels.bdLatexPanel.latex = {
        '\textbf{Wille \& Baker (1992) Example 3}';
        '';
        'Delay Differential Equations';
        '$a\,\dot y_1(t) = y_1(t-\tau_1)$';
        '$b\,\dot y_2(t) = y_1(t-\tau_1) + y_2(t-\tau_2)$';
        '$c\,\dot y_3(t) = y_2(t)$';
        'where';
        '$y_1(t), y_2(t), y_3(t)$ are state variables,';
        '$\tau_1,\tau_2$ are constant time delays.';
        '$a,b,c$ are time scale constants,';
        'Initial conditions are constant for $t{<}0$';

    % Other panels
    sys.panels.bdTimePortrait = [];    % Time Portrait
    sys.panels.bdPhasePortrait = [];   % Phase Portrait
    sys.panels.bdSolverPanel = [];     % Solver Panel
    };
end

% The DDE function. Y and dYdt are (3x1), Z is (3x2).
function dYdt = ddefun(t,Y,Z,a,b,c)
    Ylag1 = Z(:,1);                    % Y(t-tau1)
    Ylag2 = Z(:,2);                    % Y(t-tau2)
    dy1dt = Ylag1(1) ./a;              % equation (3.1)
    dy2dt = (Ylag1(1) + Ylag2(2)) ./b; % equation (3.2)
    dy3dt = Y(2) ./c;                  % equation (3.3)
    dYdt = [dy1dt; dy2dt; dy3dt];      % return vector
end
```

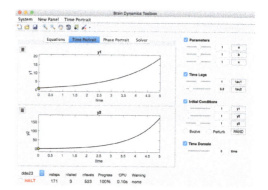

Fig. 3.1 Screenshot of the of Willé and Baker [64] model defined by Listing 3.2.

3.2 The DDE function

The DDE function defines the equations to be solved. It must have the syntax

```
function dYdt = ddefun(t,Y,Z,a,b,c,...)
```

where `Y` and `dYdt` are both $n \times 1$ vectors and `t` is a scalar, as is the case for ODEs. Parameter `Z` is an $n \times k$ matrix that contains time-lagged versions of $Y(t - \tau_i)$ in each column. As with ODEs, the model-specific parameters `a, b, c, ...` may be scalars, vectors or matrices. The DDE function for the Willé-Baker [64] model (Listing 3.2; lines 41–49) is as follows

```
function dYdt = ddefun(t,Y,Z,a,b,c)
    Ylag1 = Z(:,1);
    Ylag2 = Z(:,2);
    dy1dt = Ylag1(1)./a;
    dy2dt = (Ylag1(1) + Ylag2(2))./b;
    dy3dt = Y(2)./c;
    dYdt = [dy1dt; dy2dt; dy3dt];
end
```

where the input parameters

$$Y = \begin{bmatrix} y_1(t) \\ y_2(t) \\ y_3(t) \end{bmatrix}$$

and

$$Z = \begin{bmatrix} y_1(t-\tau_1) & y_1(t-\tau_2) \\ y_2(t-\tau_1) & y_2(t-\tau_2) \\ y_3(t-\tau_1) & y_3(t-\tau_2) \end{bmatrix}$$

are the instantaneous and time-delayed states of the system of equations. The output of the function is the column vector,

$$
\mathrm{dYdt} = \begin{bmatrix} \dot{y}_1(t) \\ \dot{y}_2(t) \\ \dot{y}_3(t) \end{bmatrix} .
$$

All of these data structures have the same number of rows. That number is determined by the combined size of all `value` fields in the `vardef` structure (lines 14–17). Similarly, the sizes of each of the model-specific parameters (a, b, c) are determined by the sizes of their respective `value` fields in the `pardef` data structure (lines 5–8). In this case, the model-specific parameters all happen to be scalars.

3.3 The DDE parameters

The `pardef` data structure defines the number and size of DDE parameters in the same manner as for ODEs. Our variant of the Willé-Baker [64] model (Listing 3.3; lines 5–8) has three scalar parameters (a=1, b=1, c=1) whose contents can be changed via the control panel of the graphic user interface. The toolbox passes those values to the DDE solver which in turn passes them onto the user-defined DDE function (lines 41–49).

3.4 The DDE variables

The `vardef` data structure defines the DDE state variables. Their names and sizes dictate how they appear in the graphic user interface. Nonetheless the contents of those variables are concatenated into a single monolithic column vector when passed to the solver — and subsequently to the user-defined DDE function. For example, the three state variables (y1, y2, y3) in Listing 3.3 are each defined as scalar values (lines 14–17) yet they are passed to the DDE function (lines 41-49) as the single column vector Y=[y1;y2;y3].

3.5 The DDE time lag parameters

Time lags are defined by the `lagdef` data structure which has the same format as the `pardef` and `vardef` data structures. Each lag parameter may be defined as a scalar, vector or matrix value however this only effects how they are displayed in the graphic user interface. The values themselves are actually passed to the solver as a single monolithic `lags` vector.

```
lags = bdGetValues(sys.lagdef)
```

The ordering of the columns in the Z matrix matches the entries in the `lags` vector. Thus the following lag definition,

```
sys.lagdef = [
    struct('name','lag1', 'value',[1 2])
    struct('name','lag2', 'value',3 )
    struct('name','lag3', 'value',[4 5 ; 6 7])
    ];
```

produces a lag vector with seven values,

```
lags = [1 2 3 4 5 6 7]'
```

and a corresponding Z matrix with seven columns,

$$Z = \begin{bmatrix} y_1(t-1) & y_1(t-2) & \ldots & y_1(t-7) \\ \vdots & \vdots & & \vdots \\ y_n(t-1) & y_n(t-2) & \ldots & y_n(t-7) \end{bmatrix}.$$

Lags are slow

It is advisable to use lag parameters sparingly because having too many lags will drastically impair the performance of the DDE solver.

3.6 DDE solver options

The format of the DDE solver option structure (`ddeoption`) is identical to that used by the MATLAB `ddeset` and `ddeget` routines. Those functions can thus be used to configured the solver options but it is usually simpler to directly assign an option by its field name.

```
% example DDE options
sys.ddeoption.AbsTol = 1e-6;
sys.ddeoption.RelTol = 1e-3;
```

A complete list of DDE solver options is given on page 39.

3.7 System structure for DDEs

This section provides a complete list of all `sys` fields that are relevant to DDEs. See Chapter 5 for fields related to display panels.

`sys.ddefun = @(t,Y,Z,a,b,c,...)`

A handle to a user-defined DDE function of the form

`dYdt = myfun(t,Y,Z,a,b,c,...)`

where t is a scalar and both dYdt and Y are $n \times 1$ vectors. Parameter Z is a $n \times k$ matrix that contains time-lagged versions of $Y(t - \tau_i)$ in each column. The user-defined parameters a, b, c, ... may be scalars, vectors or matrices. The function defines the right-hand side of the DDE. It has the same syntax required by the MATLAB DDE solvers (e.g. dde23).

`sys.pardef`

An array of structures that define the names and values of each DDE parameter in the user-defined DDE. There must be one array element for each user-defined parameter (a,b,c,...) in ddefun. The structure has the following fields:

`name = <string>`

> The name of the DDE parameter as it appears in the graphical interface. It is also used by the bdGetValue and bdSetValue functions to read and write parameter values by name (Chapter 7).

`value = <numeric>`

> The default value of the DDE parameter. It may be either a scalar, vector or matrix provided that its size matches that of the corresponding input parameter of ddefun.

`lim = [lo hi]`

> The lower and upper limits of the DDE parameter (optional). These limits are applied to the plot axes in the display panels as well as the sliders in the control panel. If no limits are specified then they are inferred from the default value(s) of the DDE parameter.

`sys.lagdef`

An array of structures defining the names and values of the lag parameters in the user-defined DDE. Multiple lag parameters can be combined as vectors or represented separately as individual scalar parameters. The total number of lags corresponds to the number of columns in the Z parameter of ddefun.

`name = <string>`

> The name of the lag parameter as it appears in the graphical interface. It is also used by the bdGetValue and bdSetValue functions to read and write parameter values by name (Chapter 7).

`value = <numeric>`

> The default value of the lag parameter (optional). It may be either a scalar, vector or matrix but all values must be non-negative.

```
lim = [lo hi]
```
The lower and upper limits of the lag parameter. These limits are applied to the plot axes in the display panels as well as the sliders in the control panel. If no limits are specified then they are inferred from the default value(s) of the lag parameter.

sys.vardef

An array of structures that define the names and values of each DDE state variable. Each structure has the following fields:

```
name = <string>
```
The name of the DDE state variable as it appears in the graphical interface. The name is also used by the `bdGetValue` and `bdSetValue` functions (Chapter 7) which can be used in scripts to read and write initial values by name.

```
value = <numeric>
```
The initial value of the DDE state variable. The numeric value may be formatted as either a scalar, vector or matrix.

```
lim = [lo hi]
```
The lower and upper limits of the state variable (optional). These limits are applied to the plot axes in the display panels as well as the sliders in the control panel. If no limits are specified then they are estimated from the initial values of the state variable.

```
solindx = <integer>
```
An array of indices that map each element of the state variable to the matching rows of the solution data (`sol.y`) produced by the solver. This field is initialised by `bdGUI` whenever it loads a system structure so there is no need to define it explicitly. It is used by display panels to extract the relevant solution data for plotting.

sys.panels

Struct containing display panel options (Chapter 5).

sys.tspan = [t0 t1]

The time span of the numerical integration where $t_0 < t_1$. This field is optional and defaults to $t_0 = 0$ and $t_1 = 1$ when omitted.

sys.tval = t0

The value of the time slider where $t0 \leq tval \leq t1$ (optional).

sys.tstep = 1

The time step used by the interpolator (optional).

sys.ddesolver = {@dde23}

A cell array of function handles to the relevant DDE solvers (optional). Note: The obsolete dde23a solver was removed in version 2023a.

sys.ddeoption

A structure containing the DDE solver options (optional). It uses the same format as the MATLAB ddeset and ddeget functions. Those options and their default values are summarised here. See the ddeset documentation for a complete description.

RelTol = 1e-3
: Error tolerance relative to the magnitude of each solution component. It controls the correct number of digits in the solution except where the solution is smaller than AbsTol.

AbsTol = 1e-6
: Absolute tolerance of the solver error. The precision of solution components below this tolerance threshold is not considered to be important.

NormControl='off'
: Use the norm of the solution rather than the absolute value of each component when computing the relative tolerance of the solver error.

OutputFcn
: Handle to a user-supplied function that is called at each time step. Available with bdSolve but not bdGUI.

OutputSel
: Specifies which solution components are passed to OutputFcn. Available with bdSolve but not bdGUI.

Stats
: Solver statistics. Available with bdSolve but not bdGUI.

InitialStep
: Sets an upper bound on the initial step taken by the solver.

MaxStep
: Sets an upper bound on each step taken by the solver.

Events
: Handle to a user-supplied function that detects points of interest in the solution.

Jumps
: Location of discontinuities.

initialY
: Initial value of the solution.

```
sys.halt = false
sys.evolve = false
sys.perturb = false
```

The initial states of the HALT, EVOLVE, PERTURB buttons.

sys.UserData

This field is reserved for user-defined data. It can be used to store model-specific function handles and data structures directly in the system structure. The contents of this field are not modified by the toolbox and there are no restrictions on what can be stored in it.

sys.auxfun = @(t,Y,a,b,c,...)

This field was deprecated in version 2018a.

sys.self = @()

This field was deprecated in version 2018a.

Chapter 4
Stochastic Differential Equations

Stochastic Differential Equations (SDEs) have the general form,

$$dY = F(t,Y)\,dt + G(t,Y)\,dW(t) \qquad (4.1)$$

where $F(t,Y)\,dt$ is the deterministic part of the system and $G(t,Y)\,dW(t)$ is the stochastic part. The dt term represents an infinitesimal time step and the $dW(t)$ term represents an infinitesimal random step. The random steps are drawn from a normal distribution with zero mean and variance dt in the limit of vanishingly small dt. This corresponds to a continuous stochastic process $W(t)$ known as a Wiener (or Brownian) noise process [18]. For a system with n state variables and m independent noise processes, the general form is best described in matrix notation,

$$\begin{bmatrix} dY_1 \\ \vdots \\ dY_n \end{bmatrix} = \begin{bmatrix} f_1 \\ \vdots \\ f_n \end{bmatrix} dt + \begin{bmatrix} g_{1,1} & \cdots & g_{1,m} \\ \vdots & & \vdots \\ g_{n,1} & \cdots & g_{n,m} \end{bmatrix} \begin{bmatrix} dW_1 \\ \vdots \\ dW_m \end{bmatrix}$$

where $[f_i] = F(t,Y)$ are the coefficients of the deterministic process and $[g_{i,j}] = G(t,Y)$ are the coefficients of the noise process, any of which may vary in time. The coefficients of the noise are arranged as a matrix so that any mixture of noise processes can be defined for each state variable.

In numerical simulations, the infinitesimal terms dt and $dW(t)$ are approximated by finite increments Δt and $\Delta W(t)$. The solvers provided with the *Brain Dynamics Toolbox* automatically generate the random increments $\Delta W(t) = N(0, \Delta t)$ for a time domain that is nominated by the user at run-time. The particular sequence of random increments that is generated is called a *realisation* of the random process. The choice of Δt is opaque to the model hence the toolbox requires that $F(t,Y)$ and $G(t,Y)$ be implemented as separate functions, as described in the next section.

4.1 Defining an SDE

We demonstrate the method of defining an SDE in the toolbox by implementing a system of n independent Ornstein-Uhlenbeck processes,

$$dy_i = \theta \left(\mu - y_i \right) dt + \sigma \, dW_i \quad \text{for} \quad i = 1 \dots n, \tag{4.2}$$

where $y_i(t)$ are the state variables and $W_i(t)$ are Wiener noise processes. The coefficients μ and θ respectively dictate the long-term mean and the rate of convergence of the noise-free system. The coefficient σ dictates the magnitude of the noise.

We model n independent processes because it is convenient to compare multiple realisations of the noise process in the same simulation (Figure 4.1) even though those processes do not interact. The system definition is shown in Listing 4.1. It corresponds to the *OrnsteinUhlenbeck* example provided with the toolbox (Table 1.1). The stochastic equation (4.2) is defined by two functions (lines 40–48) where the deterministic part is defined by sdeF and the stochastic part is defined by sdeG. The relevant SDE solvers (lines 14–15) are explicitly defined as the Euler-Maruyama method (@sdeEM) and the Stratonovich-Huen method (@sdeSH). This setting could have been omitted since those are the default solvers anyway. In any event, the solver options (lines 17–19) defining the step size (InitialStep) and the number of noise sources (NoiseSources) are mandatory for these solvers. The system parameters (lines 6–9), state variables (lines 11–12) and display panels (lines 21–37) are all defined using the same conventions previously described for ODEs (Chapter 2).

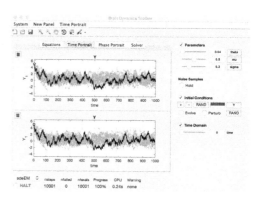

Fig. 4.1 Screenshot of $n = 20$ Ornstein-Uhlenbeck noise processes from Listing 4.1.

Listing 4.1 System definition for n independent Ornstein-Uhlenbeck processes.

```matlab
function sys = OrnsteinUhlenbeck(n)
    % SDE function handles
    sys.sdeF = @sdeF;         % deterministic part of the SDE
    sys.sdeG = @sdeG;         % stochastic part of the SDE

    % SDE parameters
    sys.pardef = [ struct('name','theta', 'value',1.0);
                   struct('name','mu',    'value',0.5);
                   struct('name','sigma', 'value',0.5) ];

    % SDE state variables
    sys.vardef = struct('name','Y',   'value',5*ones(n,1));

    % SDE solvers and options
    sys.sdesolver = {@sdeEM,@sdeSH};      % Relevant solvers
    sys.sdeoption.InitialStep = 0.01;     % Solver step size
    sys.sdeoption.NoiseSources = n;       % n noise sources

    % Latex (Equations) panel
    sys.panels.bdLatexPanel.title = 'Equations';
    sys.panels.bdLatexPanel.latex = {
        '\textbf{Ornstein-Uhlenbeck}';
        '';
        'Independent Ornstein-Uhlenbeck processes';
        '\qquad $dY_i = \theta(\mu-Y_i)dt + \sigma dW_i$';
        'where';
        '\qquad $Y_i(t)$ are the $n$ state variables,';
        '\qquad $\mu$ is the long-term mean of $Y_i(t)$,';
        '\qquad $\theta>0$ is rate of convergence,';
        '\qquad $\sigma>0$ is the volatility.' };

    % Other panels
    sys.panels.bdTimePortrait = [];       % Time Portrait
    sys.panels.bdPhasePortrait = [];      % Phase Portrait
    sys.panels.bdSolverPanel = [];        % Solver Panel
end

% The deterministic part of the equation.
function F = sdeF(t,Y,theta,mu,sigma)
    F = theta .* (mu - Y);
end

% The stochastic part of the equation.
function G = sdeG(t,Y,theta,mu,sigma)
    G = sigma .* eye(numel(Y));
end
```

Itô versus Stratonovich Calculus

The toolbox includes specialised solvers that interpret equation (4.1) according to either Itô calculus or Stratonovich calculus. These two types of stochastic calculus derive from different conceptualizations of the integral of a Wiener noise process in the limit of infinitesimally small time steps [1, 18, 32, 63]. Itô calculus defines the integral as

$$\int_0^t G(t)\, dW(t) = \lim_{N\to\infty} \sum_{n=0}^{N-1} G(t_n)\Big(W(t_{n+1}) - W(t_n)\Big)$$

where $G(t)$ is some time-varying coefficient and $W(t)$ is the Wiener noise process. It evaluates $G(t_n)$ at the beginning of the nth time interval where $t_n = n\Delta t$. In contrast, the Stratonovich calculus defines the integral as

$$\int_0^t G(t) \circ dW(t) = \lim_{N\to\infty} \sum_{n=0}^{N-1} \frac{G(t_n) + \Delta t\, G(t_{n+1})}{2} \Big(W(t_{n+1}) - W(t_n)\Big)$$

which evaluates $G(t_n)$ and $G(t_{n+1})$ at the endpoints of the nth time interval[1].

The two integrals are equally admissable but have different limits that lead to different versions of the Fokker-Planck equation [56, 62]. The Itô integral leads to

$$\partial \rho(Y,t) = -\partial_Y F \rho(Y,t) + \frac{1}{2}\partial_Y^2 G^2 \rho(Y,t)$$

whereas the Stratonovich integral leads to

$$\partial \rho(Y,t) = -\partial_Y \left(F + \frac{1}{2}\, G\, \partial_Y G\right)\rho(Y,t) + \frac{1}{2}\partial_Y^2 G^2 \rho(Y,t)$$

where $\rho(Y,t)$ is the time-dependent probability density. The deterministic parts of these equations differ by $\frac{1}{2}G(t,Y)\partial_Y G(t,Y)$ hence the two interpretations of the stochastic equation (4.1) require different numerical algorithms.

The toolbox includes the Euler-Maruyama [42] method (sdeEM) for solving systems according to the rules of Itô calculus. And it includes the Stratonovich-Heun [51] method (sdeSH) for solving systems according to the rules of Stratonovich calculus. In practice, both methods give identical results when the noise coefficients G are independent of Y (called *additive* noise). However, the two methods can produce strikingly different results when G depends on Y (called *multiplicative* noise).

We illustrate this point using the stochastic differential equation,

$$dy = -(a + b^2 y)(1 - y^2)\, dt + b(1 - y^2)\, dW, \qquad (4.3)$$

---

[1] The ∘ notation distinguishes the Stratonovich integral from the Itô integral.

from example (4.46) of Kloeden and Platen [34]. This model is included in
the toolbox under the name *KloedenPlaten446* (Table 1.1). The stochastic
differential equation (4.3) has the explicit solution,

$$y = \frac{(1+y_0)\exp(-2at+2bW_t)+y_0-1}{(1+y_0)\exp(-2at+2bW_t)-y_0+1} \tag{4.4}$$

when interpreted with Itô calculus [34]. However that solution does not apply
under the Stratonovich interpretation because the noise coefficient $b(1-y^2)$
depends on y. This is confirmed by numerical simulation where the solution computed with the Euler-Maruyama method (black line in Figure 4.2)
corresponds exactly to the explicit solution (4.4) for the Itô interpretation.
Nonetheless it differs markedly from the Stratonovich interpretation computed by the Stratonovich-Heun method (grey line in Figure 4.2). The two
results are both correct (and validly so) because they represent alternative
interpretations of the continuous stochastic equation (4.3). Thus a numerical
solution to a stochastic differential equation with multiplicative noise terms
is incomplete unless it includes a statement of which stochastic calculus was
used to interpret the equation.

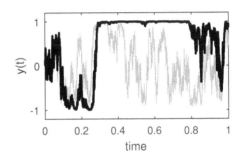

Fig. 4.2 Differing solutions
to equation (4.3) using the Itô
interpretation (black) and the
Stratonovich interpretation
(grey). The results diverge
dramatically even though the
parameters $(a=1, b=6)$, step
size $(\Delta t=0.00001)$, initial
conditions and noise samples
are identical.

4.2 The SDE functions

The functions $F(t,Y)$ and $G(t,Y)$ are defined with the same syntax,

```
function F = sdeF(t,Y,a,b,c,...)
```

and

```
function G = sdeG(t,Y,a,b,c,...)
```

where t is a scalar and Y is an $n \times 1$ vector of state variables. The model-
specific parameters a,b,c, ... may be scalars, vectors or matrices according to the contents of the value fields of the pardef data structure

(lines 6–9 of Listing 4.1). The two functions receive identical input parameters from the solver but the format of their outputs differ. The `sdeF` function should return an $n \times 1$ column vector whereas the `sdeG` function should return an $n \times m$ matrix.

In the case of our Ornstein-Uhlenbeck example (Listing 4.1) the deterministic part of the system, $F(y_i, t) = \theta(\mu - y_i)$, is implemented as

```
function F = sdeF(t,Y,theta,mu,sigma)
    F = theta .* (mu - Y);
end
```

where Y and F are both $n \times 1$ vectors. The model-specific parameters, `theta`, `mu` and `sigma`, are all scalars in this case. Similarly, the stochastic part of the system, $G(y_i, t) = \sigma$, is implemented as

```
function G = sdeG(t,Y,theta,mu,sigma)
    G = sigma .* eye(numel(Y));
end
```

where the output parameter G is an $n \times m$ matrix. It happens to be a square matrix in this case,

$$G = \begin{bmatrix} \sigma & & 0 \\ & \ddots & \\ 0 & & \sigma \end{bmatrix},$$

because the number of noise sources (m) equals the number of state variables (n). The off-diagonal entries are zero because each noise source drives each variable independently. That is usually not the case for most models.

4.3 The SDE parameters and variables

The parameters and variables of a SDE are defined using the same `pardef` and `vardef` data structures described for ODEs in Chapters 2.3–2.4.

4.4 SDE solver options

The SDE solver options are passed to the toolbox via the `sdeoption` field of the system structure (lines 17–19 in Listing 4.1). The `sdeEM` and `sdeSH` solvers are both fixed-step algorithms. The size of the time step (Δt) is defined by the `InitialStep` option. It is mandatory except when pre-generated realisations of the noise have been given in the `randn` option —

in which case the step-size is inferred from the number of noise samples. Pre-generated noise is described further in the next section. Another mandatory option for SDEs is the number of noise sources (NoiseSources) which must match the number of columns in the matrix returned by sdeG.

4.5 User-generated realisation of the noise

Ordinarily, the SDE solver generates new realisations of the noise increments $\Delta W(t)$ for every simulation run. Yet it is often advantageous to re-run a simulation with the same realisation of the noise process. The graphical user interface includes a *hold* button that freezes the current realisation of the noise for that purpose. The toolbox also allows the user to supply a pre-generated set of stochastic increments via the randn solver option.

```
sys.sdeoption.randn = randn(m,s);
```

The stochastic increments should be drawn from the standard normal distribution using the randn(m, s) function where m is the number of noise sources in the model and s is the desired number of time steps in the simulation. The SDE solver ignores the InitialStep option when the randn option is given. It instead computes the step size as

$$\Delta t = \frac{t_1 - t_0}{s - 1}$$

where the time span of the integration is tspan=[t0 t1].

4.6 System structure for SDEs

This section provides a complete list of all sys fields that are relevant to SDEs. See Chapter 5 for fields related to display panels.

sys.sdeF = @(t,Y,a,b,c,...)
   A handle to a user-defined function of the form
       F = sdeF(t,Y,a,b,c,...)
   where t is a scalar and both F and Y are n×1 vectors. The function returns the deterministic coefficients $F(t,Y)$ of the SDE as an n×1 vector. The user-defined parameters a, b, c, ... may be scalars, vectors or matrices.

sys.sdeG = @(t,Y,a,b,c,...)
   A handle to a user-defined function of the form

```
G = sdeG(t,Y,a,b,c,...).
```
The function returns the noise coefficients $G(t,Y)$ of the SDE as an $m \times n$ matrix where m is the number of noise noises and n is the number of state variables. The input parameters are identical to `sdeF` (above).

sys.pardef

An array of structures that define the names and values of each parameter in the user-defined SDE. There must be one array element for each user-defined parameter (`a,b,c,...`) expected identically by both `sdeF` and `sdeG`. The structure has the following fields:

```
name = <string>
```
The name of the SDE parameter as it appears in the graphical interface. It is also used by the `bdGetValue` and `bdSetValue` functions to read and write parameter values by name (Chapter 7).

```
value = <numeric>
```
The default value of the SDE parameter. It may be either a scalar, vector or matrix provided that its size matches that of the corresponding input parameter of `sdeF` and `sdeG`.

```
lim = [lo hi]
```
The lower and upper limits of the SDE parameter (optional). These limits are applied to the plot axes in the display panels as well as the sliders in the control panel. If no limits are specified then they are inferred from the default value(s) of the parameter.

sys.vardef

An array of structures that define the names and values of each SDE state variable. Each structure has the following fields:

```
name = <string>
```
The name of the SDE state variable as it appears in the graphical interface. The name is also used by the `bdGetValue` and `bdSetValue` functions (Chapter 7) which can be used in scripts to read and write initial values by name.

```
value = <numeric>
```
The initial value of the SDE state variable. The numeric value may be formatted as either a scalar, vector or matrix.

```
lim = [lo hi]
```
The lower and upper limits of the state variable (optional). These limits are applied to the plot axes in the display panels as well as the sliders in the control panel. If no limits are specified then they are estimated from the initial values of the state variable.

```
solindx = <integer>
```
> An array of indices that map each element of the state variable to the matching rows of the solution data (sol.y) produced by the solver. This field is initialised by bdGUI whenever it loads a system structure so there is no need to define it explicitly. It is used by display panels to extract the relevant solution data for plotting.

sys.panels

> A structure containing the various display panel options. The fields within this structure correspond to the names of the display panel classes. The options for each display panel are described in Chapter 5.

sys.tspan = [t0 t1]

> The time span of the numerical integration where $t_0 < t_1$. This field is optional and defaults to $t_0=0$ and $t_1=1$ when omitted.

sys.tval = t0

> The value of the time slider where t0 $\leq$ tval $\leq$ t1 (optional).

sys.tstep = 1

> The time step used by the solver (optional). It defines the value of the InitialStep option used by the SDE solvers.

sys.sdesolver = {@sdeEM, @sdeSH}

> A cell array of function handles to the relevant SDE solvers (optional). The sdeEM solver implements the Euler-Maruyama method. The sdeSH solver implements the Stratonovich-Heun method.

sys.sdeoption

> A structure containing the SDE solver options. The fields are as follows:
>
> ```
> InitialStep
> ```
> > The fixed step size taken by the solver.
>
> ```
> NoiseSources = n
> ```
> > The number of Wiener processes (mandatory).
>
> ```
> randn = randn(n,s)
> ```
> > Optional user-supplied Wiener noise samples where n is the number of Wiener processes and s is the number of time points. If this option is supplied then the value n must match NoiseSources.

sys.halt = false
sys.evolve = false

`sys.perturb = false`

The initial states of the HALT, EVOLVE, PERTURB buttons.

`sys.UserData`

This field is reserved for user-defined data. It can be used to store model-specific function handles and data structures directly in the system structure. The contents of this field are not modified by the toolbox and there are no restrictions on what can be stored in it.

`sys.auxfun = @(t,Y,a,b,c,...)`

This field was deprecated in version 2018a.

`sys.self = @()`

This field was deprecated in version 2018a.

Chapter 5
Display Panels

Display panels are essentially plug-in modules that can be loaded into `bdGUI` at run-time. Each one offers a different visualisation of the computed solution. The toolbox ships with a collection of display panel classes in the *bdtoolkit/panels* directory which are included in the *New Panel* toolbar menu. User-defined panels can also be included in the menu by explicitly naming them in the `panels` section of the model's system structure. All display panels must be findable on the MATLAB search path otherwise they will fail to load. Panels that are explicitly named in the system structure will be automatically loaded at start-up.

```
% Automatically load the Time Portrait and
% Phase Portrait panels at system start-up
sys.panels.bdTimePortrait = [];
sys.panels.bdPhasePortrait = [];
```

The field name must match the name of the panel class. The field itself can be empty or it can include additional configuration options for the panel.

```
% Configuration options for the Time Portrait
sys.panels.bdTimePortrait.title = 'Time Portrait';
sys.panels.bdTimePortrait.transients = 'on';
sys.panels.bdTimePortrait.markers = 'on';
sys.panels.bdTimePortrait.modulo = 'off';
```

The options are specific to each panel class. In practice, the default configurations are usually adequate. A notable exception is the LaTeX Equations panel (`bdLatexPanel`) which requires model-specific LaTeX strings for rendering the mathematical equations.

5.1 LaTeX Equations

The Equations panel (Figure 5.1) renders mathematical equations using MATLAB's built-in LaTeX interpreter [38]. The LaTeX strings are defined in the `latex` field of the panel's configurations options (Table 5.1).

```
sys.panels.bdLatexPanel.latex =
    {string1, string2, ... , stringN};
```

Each string corresponds to one line of typesetting. For example,

```
string1 = 'The equation $y=ax+b$ is linear.';
```

where mathematical symbols are delimited by $...$. Empty strings produce a vertical space. The LaTeX commands can be edited interactively by toggling the *Edit* menu option. The edited strings are saved in the model's system structure. See [59, 36] for an introduction to LaTeX typesetting.

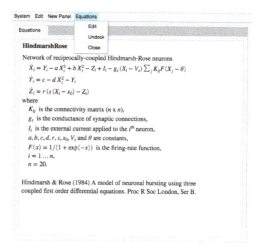

Fig. 5.1 LaTeX Equations for the `HindmarshRose` model. The underlying LaTeX commands can be modified using the *Edit* menu.

Table 5.1 Configuration options for `sys.panels.bdLatexPanel`

Option	Description
`title = 'Equations'`	Title text for the panel.
`latex = {str1, str2, ... }`	Cell array of LaTeX strings.
`fontsize = 16`	The font size (in pixels).

5.2 Time Portrait

The Time Portrait (Figure 5.2) displays the time course of two state variables. Multi-variate data are plotted as light grey traces with one trajectory emphasized in heavy black. The subscript of the highlighted trajectory can be changed by toggling the checkbox next to the selector. The transient part of the trajectory can be hidden by toggling the *Transients* menu option. The plot limits are defined by the upper and lower limits of the initial conditions in the control panel. The *Calibrate* menu adjusts those limits to fit the data. The *Modulo* option wraps the plot lines at the boundaries. The *Auto Steps* option plots the exact time steps taken by the solver, otherwise the trajectories are interpolated using fixed time steps. The system configuration options for the Time Portrait are listed in Table 5.2.

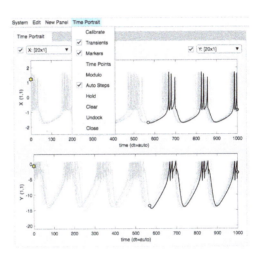

Fig. 5.2 Time Portrait for a ring of $n=20$ Hindmarsh-Rose neurons [28]. The state variables (x and y) are both 20×1 vectors. The time courses of the selected neuron, $x_1(t)$ and $y_1(t)$, are plotted in black. The other neurons are plotted as background traces. The yellow hexagon marks the initial conditions. The filled circle marks the end of the trajectory. The open circle marks the position of the time slider.

Table 5.2 Configuration options for `sys.panels.bdTimePortrait`

Option	Description
`title = 'Time Portrait'`	Title text for the panel.
`transients = 'on'`	Show the transient part of the trajectory.
`markers = 'on'`	Show the markers.
`points = 'off'`	Show the discrete points.
`modulo = 'off'`	Wrap the plot lines at the boundaries.
`autostep = 'on'`	Use the time steps chosen by the solver.
`hold = 'off'`	Hold all plots.
`selector1 = {[1],[1],[1]}`	Selector item, row subscript, col subscript.
`selector2 = {[1],[1],[1]}`	Selector item, row subscript, col subscript.

5.3 Time Cylinder

The Time Cylinder (Figure 5.3) maps the time course of a state variable onto cylindrical coordinates. It is useful for visualising phase variables. The polar coordinate spans the limits of the state variable, as defined in the control panel (Initial Conditions). The transient part of the trajectory can be hidden from view by toggling the *Transients* menu option. The *Auto Steps* option plots the exact time steps taken by the solver. The *Relative Phases* option rotates the phases at each time point so that the first element of a multi-variate signal is always pinned at zero. The resulting phase angles are accessible via the workspace interface (gui.panels.bdTimeCylinder.theta). The system configuration options for the Time Cylinder are listed in Table 5.3.

Fig. 5.3 Time Cylinder for a multi-variate time series X with $n=20$ elements. The phase of X_1 is pinned at zero. The highlighted trajectory is X_{11}. The yellow hexagram marks the initial conditions. The filled circle marks the end point. The open circle marks the position of the time slider. The large rotator knob controls the roll angle of the cylinder mesh. The slider controls its azimuth angle. The smaller rotator knob controls the opacity of the cylinder.

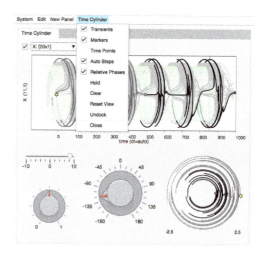

Table 5.3 Configuration options for sys.panels.bdTimeCylinder

Option	Description
title = 'Time Cylinder'	Title text of the panel.
transients = 'on'	Show the transient part of the trajectory.
markers = 'on'	Show the markers.
points = 'off'	Show the discrete points.
relphase = 'off'	Show relative phases.
azimuth = 0	Azimuth (yaw) of the cylinder mesh.
elevation = 0	Elevation (roll) of the cylinder mesh.
opacity = 0	Opacity of the cylinder mesh.
selector = {[1],[1],[1]}	Selector item, row subscript, col subscript.

5.4 Phase Portrait (2D)

The 2D Phase Portrait (Figure 5.4) plots the relationship between two selected state variables. For multi-variate data, the subscripts of the selected variables can be chosen by toggling the checkbox next to the selectors. The *Calibrate* menu adjusts the plot limits to fit the data. The transient part of the trajectory can be hidden by toggling the *Transients* menu. The *Modulo* menu wraps the plot lines at the boundaries. The *Auto Steps* menu plots the exact time steps taken by the solver. Interpolated time points are plotted otherwise. The system configuration options for the Phase Portrait are listed in Table 5.4. See Section 5.5 for phase portraits with three state variables.

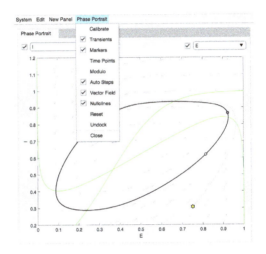

Fig. 5.4 Phase Portrait of the Wilson-Cowan neural mass model [65]. The trajectory converges onto a stable limit cycle (heavy black line). The yellow hexagon marks the initial conditions. The filled circle marks the end point. The open circle marks the position of the time slider. The vector field shows the direction of flow. Nullclines are shown in green. An equilibrium point exists where the nullclines intersect. In this case it is unstable.

Table 5.4 Configuration options for sys.panels.bdPhasePortrait

Option	Description
title = 'Phase Portrait 2D'	Title text for the panel.
transients = 'on'	Show the transient part of the trajectory.
markers = 'on'	Show the markers.
points = 'off'	Show the discrete points.
modulo = 'off'	Wrap the plot lines at the boundaries.
autostep = 'on'	Use the time steps chosen by the solver.
vectorfield = 'off'	Show the vector field.
nullclines = 'off'	Show the nullclines.
hold = 'off'	Hold all plots.
selectorX = {[1],[1],[1]}	Selector item, row subscript, col subscript.
selectorY = {[2],[1],[1]}	Selector item, row subscript, col subscript.

5.5 Phase Portrait (3D)

The 3D variant of the Phase Portrait (Figure 5.5) plots the relationship between three selected state variables. For multi-variate data, the subscripts of the selected variables can be chosen by toggling the checkbox next to the selectors. The operating principles are similar to that of the 2D Phase Portrait (Section 5.4) except that vector fields and nullclines are not supported. The system configuration options for the 3D Phase Portrait are listed in Table 5.5.

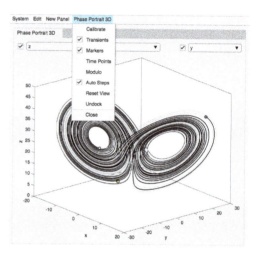

Fig. 5.5 Phase Portrait of the chaotic Lorenz attractor [40]. The yellow hexagram marks the initial conditions. The filled circle marks the end point of the trajectory. The open circle marks the position of the time slider.

Table 5.5 Configuration options for `sys.panels.bdPhasePortrait3D`

Option	Description
`title = 'Phase Portrait 3D'`	Title text for the panel.
`transients = 'on'`	Show the transient part of the trajectory.
`markers = 'on'`	Show the markers.
`points = 'off'`	Show the discrete points.
`modulo = 'off'`	Wrap the plot lines at the boundaries.
`autostep = 'on'`	Use the time steps chosen by the solver.
`hold = 'off'`	Hold all plots.
`selectorX = {[1],[1],[1]}`	Selector item, row subscript, col subscript.
`selectorY = {[2],[1],[1]}`	Selector item, row subscript, col subscript.
`selectorZ = {[3],[1],[1]}`	Selector item, row subscript, col subscript.

5.6 Solver Panel

The Solver panel (Figure 5.6) allows the auto-stepper algorithm to be tuned. The upper plot shows the time course of $\|dY\|$ which gives a general indication of how the solution changes with time. The lower plots shows the time steps (dt) taken by the solver. The auto-stepper can be tuned for efficiency by adjusting the error tolerances (AbsTol and RelTol) and stepping limits (InitialStep and MaxStep). See the MATLAB odeset documentation for the details of these solver options.

In the Brain Dynamics Toolbox, the solver options can be pre-configured in the model's odeoption, ddeoption and sdeoption fields. These fields are separate from the Solver Panel configuration options (Table 5.6) which only control the appearance of the panel. Fixed-step solvers (e.g. odeEul, sdeEM and sdeSH) use the InitialStep option to define the size of the time step. In the Solver Panel, the InitialStep automatically follows the global time step parameter (Time Domain).

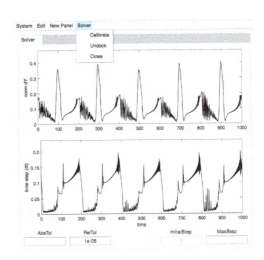

Fig. 5.6 Solver panel. The upper axes shows the progression of the vector norm of *dY* versus simulation time. The lower axes shows the size of each step versus simulation time. AbsTol and RelTol are the error tolerances of the auto-stepper algorithm. InitialStep and MaxStep control the step sizes. The InitialStep option mirrors the time step parameter of the Time Domain controls.

Table 5.6 Configuration options for sys.panels.bdSolverPanel

Option	Description
title = 'Solver'	Title text for the panel.

5.7 Space-Time Panel

The Space-Time panel (Figure 5.7) plots the time course of multi-variate data as a surface plot. It is intended for vector-based variables where the elements of the vector are treated as equally spaced nodes. The surface is constructed using the matlab `pcolor` command. It can be viewed in relief by rotating the view. The colour scale honours the limits of the state variable, as defined by the control panel. The *Calibrate* menu adjusts those limits to fit the data. The *Clipping* option truncates the surface beyond those limits. The *Modulo* option wraps the surface plot at the boundaries. The *Blend* option toggles the shading of the facets between *flat* and *interpolated* modes. The *YDir Reverse* option reverses the direction of the spatial axis. The system configuration options for the Space-Time panel are listed in Table 5.7.

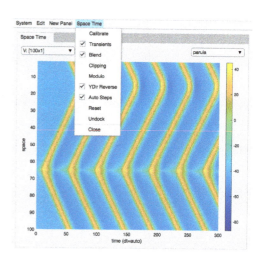

Fig. 5.7 Space-Time plot of a ring of $n=100$ Morris-Lecar neurons [43]. In this case, a constant stimulus is applied to neuron 65 which emits successive waves of activation that propagate around the ring in both directions. The waves collide at the far side of the ring (neuron 15) and annihilate one another.

Table 5.7 Configuration options for `sys.panels.bdSpaceTime`

Option	Description
`title = 'Space Time'`	Title text for the panel.
`transients = 'on'`	Show the transient part of the trajectory.
`yreverse = 'off'`	Reverse the direction of the space axis.
`blend = 'off'`	Interpolated versus flat surface shading.
`clipping = 'off'`	Clip the surface at the colour limits.
`modulo = 'off'`	Wrap the surface at the colour limits.
`colormap = 'parula'`	Name of the colour map.
`autostep = 'on'`	Use the time steps chosen by the solver.
`selector = {[1],[1],[1]}`	Selector item, row subscript, col subscript.

5.8 Space 2D Panel

The Space 2D panel (Figure 5.8) provides a snapshot of a two-dimensional ($m{\times}n$) state variable. The timing of the snapshot follows the time slider in the control panel. The colour scale honours the limits of the selected variable, as defined by the control panel. The *Calibrate* menu adjusts those limits to fit the data. The *Modulo* option wraps the colour scale at the limits. The image is a matlab primitive graphics `Image` object which does not support data cursors under the new matlab uitools interface. The image data can be accessed via the workspace interface (`gui.panels.bdSpace2D`). The system configuration options for the Space 2D panel are listed in Table 5.8.

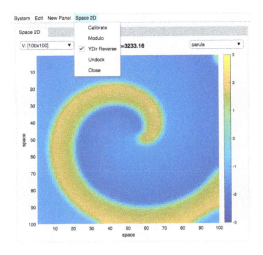

Fig. 5.8 Space 2D panel showing a rotating spiral in a sheet of $100{\times}100$ FitzHugh-Nagumo neurons [16, 44].

Table 5.8 Configuration options for `sys.panels.bdSpace2D`

Option	Description
`title = 'Space 2D'`	Title text for the panel.
`modulo = 'off'`	Wrap the surface at the colour limits.
`yreverse = 'off'`	Reverse the direction of the space axis.
`colormap = 'parula'`	Name of the colour map.
`selector = {[1],[1],[1]}`	Selector item, row subscript, col subscript.

5.9 Bifurcation Panel (2D)

The 2D Bifurcation panel (Figure 5.9) plots the trajectory of one state variable versus one system parameter. The trajectories of successive simulation runs are accumulated. The complete bifurcation diagram is thus generated incrementally while the bifurcation parameter is manipulated. Steady-state diagrams are obtained by hiding the transient dynamics via the *Transients* menu option. Forward solutions can be followed by evolving the initial conditions while ramping the bifurcation parameter. The *Perturb* option (Initial Conditions) can assist with escaping solutions that are only marginally unstable. The axes can be cleared explicitly with the *Clear* menu item. The system configuration options for the Bifurcation Panel are listed Table 5.9. See Section 5.10 for bifurcation diagrams in two state variables.

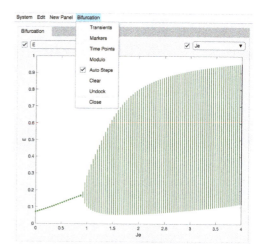

Fig. 5.9 Bifurcation diagram of the Wilson-Cowan [65] neural mass model showing the onset of oscillations in $E(t)$ at J_e=0.9. In this case the oscillations emerge via a supercritical Hopf bifurcation.

Table 5.9 Configuration options for `sys.panels.bdBifurcation`

Option	Description
`title = 'Bifurcation 2D'`	Title text for the panel.
`transients = 'on'`	Show the transient part of the trajectory.
`markers = 'on'`	Show the markers.
`points = 'off'`	Show the discrete points.
`modulo = 'off'`	Wrap the plot lines at the boundaries.
`autostep = 'on'`	Use the time steps chosen by the solver.
`selectorX = {[1],[1],[1]}`	Selector item, row subscript, col subscript.
`selectorY = {[1],[1],[1]}`	Selector item, row subscript, col subscript.

5.10 Bifurcation Panel (3D)

The 3D variant of the Bifurcation Panel (Figure 5.10) plots two selected state variables versus one system parameter. The operating principles are similar to that of the standard Bifurcation Panel (Section 5.9). The axes accumulates the trajectories of successive simulation runs until it is explicitly cleared using the *Clear* menu. Steady-state diagrams are obtained by hiding the transient dynamics with the *Transients* menu. Forward solutions can be followed by evolving the initial conditions while ramping the bifurcation parameter. The *Perturb* option (Initial Conditions) can assist with escaping solutions that are only marginally unstable. The system configuration options for the Bifurcation Panel (3D) are listed Table 5.10.

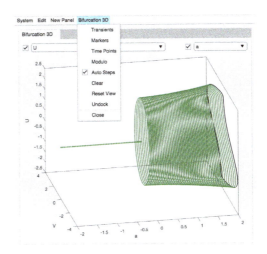

Fig. 5.10 Bifurcation diagram of the Van der Pol oscillator [61] showing the steady-state solutions, $U(t)$ and $V(t)$, for a range of parameters $a \in [-2, 2]$.

Table 5.10 Configuration options for `sys.panels.bdBifurcation3D`

Option	Description
`title = 'Bifurcation 3D'`	Title text for the panel.
`transients = 'on'`	Show the transient part of the trajectory.
`markers = 'on'`	Show the markers.
`points = 'off'`	Show the discrete points.
`modulo = 'off'`	Wrap the plot lines at the boundaries.
`autostep = 'on'`	Use the time steps chosen by the solver.
`selectorX = {[1],[1],[1]}`	Selector item, row subscript, col subscript.
`selectorY = {[2],[1],[1]}`	Selector item, row subscript, col subscript.
`selectorZ = {[1],[1],[1]}`	Selector item, row subscript, col subscript.

5.11 Jacobian Panel

The Jacobian Panel (Figure 5.11) evaluates the Jacobian of the ODE,

$$
\mathbf{J} = \begin{pmatrix} \dfrac{\partial F_1}{\partial y_1} & \cdots & \dfrac{\partial F_1}{\partial y_n} \\ \vdots & & \vdots \\ \dfrac{\partial F_n}{\partial y_1} & \cdots & \dfrac{\partial F_n}{\partial y_n} \end{pmatrix},
$$

at the point $Y(t) = [y_1(t), \ldots, y_n(t)]$ in the forward solution. The value of t is taken from the time slider in the graphical user interface. The computation is exact when the system includes an explicit definition of the Jacobian (Section 2.7) otherwise the Jacobian is estimated with finite differences. The trace and determinant of the Jacobian matrix are computed with the matlab functions of the same name (*trace* and *det*). The results are accessible to the workspace via the `gui.panels.bdDFDY` handle. The title of the panel is the only configurable option (Table 5.11). The Jacobian is undefined for DDEs and SDEs.

Fig. 5.11 Jacobian of the linear system of ODEs (2.1) with parameters $a=1$, $b=-1$, $c=10$ and $d=-2$.

Table 5.11 Configuration options for `sys.panels.bdDFDX`

Option	Description
`title = 'dFdY'`	Title text for the panel.

5.12 Eigenvalues Panel

The Eigenvalues Panel (Figure 5.12) computes the eigenvalues and eigen-vectors of the Jacobian of the ODE,

$$
J = \begin{pmatrix} \dfrac{\partial F_1}{\partial y_1} & \cdots & \dfrac{\partial F_1}{\partial y_n} \\ \vdots & & \vdots \\ \dfrac{\partial F_n}{\partial y_1} & \cdots & \dfrac{\partial F_n}{\partial y_n} \end{pmatrix},
$$

evaluated at the point $Y(t) = [y_1(t), \ldots, y_n(t)]$. The value of t is taken from the time slider in the graphical user interface. The Jacobian is computed using the same method as the Jacobian panel (Section 5.11). The eigenvalues are computed using matlab's eig function. The results are accessible to the workspace via the $gui.panels.bdEigenvalues$ handle. The compu-tations are expensive when n is large, so it is best to only apply the panel to small systems of equations. The title of the panel is the only configuration option (Table 5.12). The Jacobian is undefined for DDEs and SDEs.

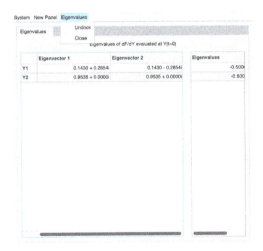

Fig. 5.12 Eigenvalues of the linear system of ODEs (2.1) evaluated at the point $Y(0) = \big(x(0), y(0)\big) = (0,0)$.

Table 5.12 Configuration options for sys.panels.bdEigenvalues

Option	Description
title = 'Eigenvalues'	Title text for the panel.

5.13 Auxiliary Panel

The Auxiliary panel (Figure 5.13) provides a blank canvas that modellers can use for custom plots without having to develop a custom display panel. The plotting is done by user-defined functions, as described in Chapter 6. Handles to those functions are assigned to the panel's `auxfun` field when the model is constructed.

```
sys.panels.bdAuxiliary.auxfun = {@auxfun1, ...
    @auxfun2, @auxfun3};
```

The `auxfun` field can contain multiple function handles, all of which appear (by name) in the Auxiliary Panel's selector menu. The chosen function is executed every time the solver generates a new solution. Apart from plotting, it can also return a custom `UserData` structure that is accessible to the workspace via `gui.panels.bdAuxiliary`. The system configuration options for the Auxiliary Panel are listed in Table 5.13.

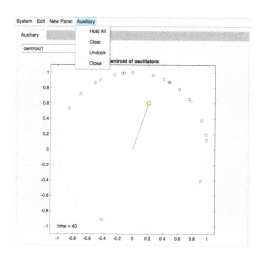

Fig. 5.13 Auxiliary panel for the `KuramotoNet` model with $n=20$ oscillators. In this case, the *centroid1* function plots the centroid of the phases of the oscillators at a given time point. The timing follows the time domain slider in the control panel. See Section 6.1 for the implementation details.

Table 5.13 Configuration options for `sys.panels.bdAuxiliary`

Option	Description
`title = 'Auxiliary'`	Title text for the panel.
`hold = 'off'`	Hold the graphical plots between runs.
`auxfun = {@fun1,@fun2,...}`	Handles to user-defined plot functions.

5.14 Correlations Panel

The Correlations panel (Figure 5.14) uses the MATLAB `corrcoef` function to compute pairwise linear correlations among multi-variate time series. The correlation is computed from the interpolated time series using equidistant time steps, as defined by the global `step` parameter (Time Domain). The transient part of the trajectory may be excluded from the computation with the *Transients* menu option. The correlation coefficients (R) and the interpolated time-series (Y) are both accessible via the workspace interface (`gui.panels.bdCorrPanel`). The correlation coefficients may also be viewed as a data table using the *Table* menu option. The system configuration options for the Correlations Panel are listed in Table 5.14.

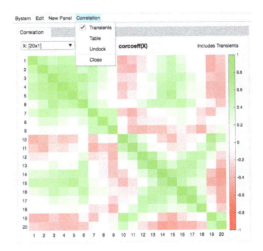

Fig. 5.14 Correlation panel. The $(i, j)^{th}$ element of the correlation matrix is the correlation between the time series in the ith and j^{th} elements of the state variable. In this case the selected variable has n=20 elements.

Table 5.14 Configuration options for `sys.panels.bdCorrPanel`

Option	Description
`title = 'Correlation'`	Title text of the panel.
`transients = 'on'`	Include the transients in the correlation.
`table = 'off'`	View the coefficients as a data table.
`selector = {[1],[1],[1]}`	Selector item, row subscript, col subscript.

5.15 Hilbert Transform

The Hilbert Transform (Figure 5.15) computes the phase of a time series from its complex-valued analytic signal [41]. The transform is computed from the interpolated time series with equidistant time steps, as defined by the global `step` parameter. The phases are then plotted in cylindrical coordinates. The transients may be excluded from the computation with the *Transients* menu option. The *Relative Phases* menu option rotates the phases at each time point so that the first element of a multi-variate signal is always pinned at zero. The analytic signal (h) and its phase angles (p) are accessible via the workspace interface (`gui.panels.bdHilbert`). The system configuration options for the Hilbert Transform panel are listed in Table 5.15.

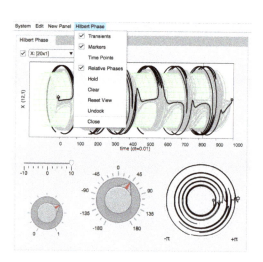

Fig. 5.15 Hilbert phases of a multi-variate time series X with $n=20$ elements. The phase of X_1 is pinned at zero. The highlighted trajectory is X_{12}. The yellow hexagram marks the initial conditions. The filled circle marks the end point. The large rotator knob controls the roll angle of the cylinder mesh. The slider controls its azimuth angle. The smaller rotator knob controls the opacity of the cylinder.

Table 5.15 Configuration options for `sys.panels.bdHilbert`

Option	Description
`title = 'Hilbert'`	Title text of the panel.
`transients = 'on'`	Include the transients.
`markers = 'on'`	Show the markers.
`points = 'off'`	Show the discrete points.
`relphase = 'off'`	Compute the relative phases.
`azimuth = 0`	Azimuth (yaw) of the cylinder mesh.
`elevation = 0`	Elevation (roll) of the cylinder mesh.
`opacity = 0`	Opacity of the cylinder mesh.
`selector = {[1],[1],[1]}`	Selector item, row subscript, col subscript.

5.16 Surrogate Data Panel

The Surrogate Data panel (Figure 5.16) generates a randomly shuffled version of a selected time series. The surrogate signal has the same amplitude distribution, autocorrelation distribution and cross-spectral density functions as the original time series but all phase information is destroyed [4]. The surrogate is regenerated automatically whenever the solver returns a new solution. It can also be regenerated manually using the *Recompute* button. The surrogate signal is computed from the interpolated signal with equidistant time steps, as defined by the global `step` parameter. The interpolated time-series (`y`) and its surrogate (`ysurr`) are both accessible via the workspace interface (`gui.panels.bdSurrogate`). The system configuration options for the Surrogate Data panel are listed in Table 5.16.

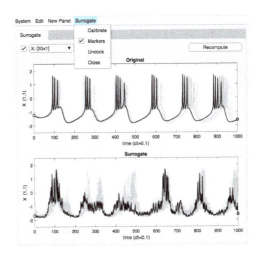

Fig. 5.16 Surrogate Data panel. The upper plot shows the original time series and the lower plot shows its phase-randomised surrogate.

Table 5.16 Configuration options for `sys.panels.bdSurrogate`

Option	Description
`title = 'Surrogate'`	The title of the panel tab.
`markers = 'on'`	Show the markers.
`selector = {[1],[1],[1]}`	Selector item, row subscript, col subscript.

5.17 System Log

The System Log (Figure 5.17) is a debugging tool for tracing the system events that are generated by the `bdSystem` class. It issues the REDRAW event whenever one or more aspects of the system has changed and needs to be redrawn. The display panels (and other instrumentation) listen for these events to manage their graphics. Information describing the properties that have changed is included in the accompanying `bdRedrawEvent` event data class. The RESPOND event is issued after the solver has computed a new solution and the display panels have redrawn themselves. It provides an opportunity for the display panels to update the `bdSystem` object in response to the newly computed solution. The PUSH event is used to notify the Undo Stack to save the current state of the system. The System Log's title string is the only configurable option in the system structure (Table 5.17).

Fig. 5.17 The System Log panel is a debugging tool for tracing system events.

Table 5.17 Configuration options for `sys.panels.bdSystemLog`

Option	Description
title = 'System Log'	Title text of the panel.

Chapter 6
Auxiliary Plot Functions

The Auxiliary panel (bdAuxiliary) provides a blank axes which is convenient for rendering model-specific plots. The plotting is implemented with a user-defined function that is called by the Auxiliary panel whenever the current solution needs replotting. That function is typically defined in the same script as the model. An Auxiliary panel can have multiple plotting functions. Handles to each one is assigned to its auxfun field when the model is constructed.

```
sys.panels.bdAuxiliary.auxfun = {@auxfun1, ...
    @auxfun2, @auxfun3};
```

The syntax of the callback functions differ slightly between the types of differential equations. For ODEs and SDEs, the callback syntax is,

```
function UserData = auxfun(ax,t,sol,a,b,c,...)
```

where ax is the handle to the panel's axes, t is the current value of the time domain slider, sol is the solution structure returned by the solver, and a,b,c,... are the parameters of the model. For ODEs, parameters a,b,c follow the syntax of odefun. For SDEs, they follow the syntax of sdeF and sdeG. For DDEs, the callback syntax is,

```
function UserData = auxfun(ax,t,sol,lag1,lag2,
    a,b,c,...)
```

where the time lag parameters lag1,lag2,... are identical to that in ddefun. Auxiliary functions may optionally return a custom UserData structure which is accessible via the workspace interface.

```
>> gui.panels.bdAuxiliary
    bdAuxiliary with properties:
        options: [1 1 struct]
           axes: [1 1 UIAxes]
       UserData: [1 1 struct]
```

6.1 Kuramoto Order Parameter

The *KuramotoNet* model provides some good examples of auxiliary plot functions. It is a weighted network of Kuramoto phase oscillators [37],

$$\frac{d\theta_i}{dt} = \omega_i + \frac{k}{n}\sum_j K_{ij}\sin(\theta_i - \theta_j)$$

where θ_i is the phase of the ith oscillator, ω_i is its natural oscillation frequency, K is an $n \times n$ network connectivity matrix, and k is a global scaling constant. The sinusoidal phase coupling term approximates the synchronisation behaviour of a broad class of neurons [47]. The Kuramoto model can thus be used to synchronization in neuronal networks [3].

A key metric for this model is the order parameter,

$$R(t) = \frac{1}{n}\left|\sum_i \exp\left(\mathbf{i}\,\theta_i(t)\right)\right|$$

where $\mathbf{i} = \sqrt{-1}$ is a complex number. Geometrically, R corresponds to the magnitude of the centroid of the oscillators after projecting them onto the unit circle in the complex plane [37]. The order parameter approaches $R=1$ when all oscillators converge to the same phase. Conversely it approaches $R=0$ when the phases are evenly distributed around the circle. The order parameter is thus an important metric of phase synchronisation.

The *KuramotoNet* model defines two auxiliary functions for plotting the centroid of the oscillators at a given instance of time (Figure 6.1). Namely,

```
function centroid1(ax,t,sol,Kij,k,omega)
function centroid2(ax,t,sol,Kij,k,omega)
```

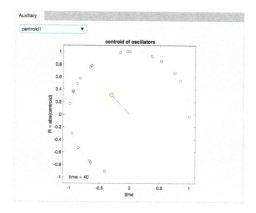

Fig. 6.1 The centroid1 auxiliary plot of the *KuramotoNet* model with $n=20$ oscillators. The oscillator phases (small circles) are plotted on the unit circle. The yellow paddle marks the centroid of the phases. Its radius defines the order parameter (R).

Listing 6.1 The `centroid1` auxiliary plot function from the *KuramotoNet* model.

```
1  function centroid1(ax,t,sol,Kij,k,omega)
2      % Get the phases of the oscillators at time t
3      theta = bdEval(sol,t);
4
5      % Project the phases into the complex plane.
6      ztheta = exp(1i.*theta);
7
8      % Plot the centroid.
9      centroidplot(ax,ztheta);
10     text(-1,-1,num2str(t,'time = %g'));
11     title('centroid of oscillators');
12 end
13
14 % Utility function for plotting the centroid
15 function centroidplot(ax,ztheta)
16     % compute the phase centroid
17     centroid = mean(ztheta);
18
19     % plot the unit circle
20     plot(ax,exp(1i.*linspace(-pi,pi,100)), ...
21         'color',[.75 .75 .75]);
22
23     % plot the oscillator phases on the unit circle
24     plot(ax,ztheta,'o','color','k');
25
26     % plot the centroid (yellow paddle)
27     plot(ax,[0 centroid], 'color', 'k');
28     plot(ax,centroid,'o','color','k', 'Marker','o', ...
29         'MarkerFaceColor','y', 'MarkerSize',10);
30
31     % axis limits etc
32     axis(ax,'equal');
33     xlim(ax,[-1.1 1.1]);
34     ylim(ax,[-1.1 1.1]);
35 end
```

where `ax` is a handle to the Auxiliary panel's axes (`UIAxes`), `t` is the current value of the time domain slider, `sol` is the current solution returned by the solver, and parameters `Kij`, `k`, and `omega` are defined by the ODE. The only difference between theses two functions is that `centroid2` uses a rotating frame which pins the first oscillator at $\theta_1 = 0$.

The source code for `centroid1` is shown in Listing 6.1. It uses the `bdEval` function (line 3) to interpolate the incoming solution (`sol`) at the given time `t`. This interpolation is necessary because the incoming `t` value is unlikely to correspond to any of the discrete time steps in the numerical solution. The instantaneous oscillator phases (theta) are then mapped to the complex plane (ztheta) and plotted (line 9) by the centroidplot utility function (line 14–35) which is shared by centroid1 and centroid2.

The *KuramotoNet* model also defines the KuramotoR auxiliary function (Listing 6.2) for plotting the time course of the order parameter (Figure 6.2). In this case, the centroid is computed at all time steps of the incoming solution (lines 2–6). The magnitude of the centroid is then plotted versus time (line 9) and the axes appropriately scaled and centred (lines 11–18).

Listing 6.2 The KuramotoR auxiliary plot function from the *KuramotoNet* model.

```
1   function KuramotoR(ax,t,sol,Kij,k,omega)
2       % Project the phases into the complex plane.
3       ztheta = exp(1i.*sol.y);
4
5       % compute the running phase centroid
6       centroid = mean(ztheta);
7
8       % plot the amplitide of the centroid versus time.
9       axis(ax,'normal');
10      plot(ax,sol.x,abs(centroid),'color','k','linewidth',1);
11
12      % axis limits etc
13      t0 = min(sol.x([1 end]));
14      t1 = max(sol.x([1 end]));
15      xlim(ax,[t0 t1]);
16      ylim(ax,[-0.1 1.1]);
17      xlabel(ax,'time');
18      ylabel(ax,'R = abs(centroid)');
19      title(ax,'Kuramoto Order Parameter (R)');
20  end
```

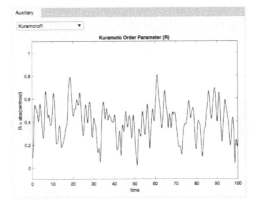

Fig. 6.2 The `KuramotoR` auxiliary function plots the time course of $R(t)$.

6.2 Hodgkin-Huxley Ionic Currents

The Hodgkin-Huxley equations [30] explain the nerve action potential in terms of voltage-gated sodium and potassium channels in the cell membrane. The change in the membrane potential (V) is described by,

$$C \frac{dV}{dt} = I - I_{Na} - I_K - I_L$$

where C is the capacitance of the membrane. The right-hand side of the equation represents the combined currents that flow across the membrane. It includes the flow of Na^+ and K^+ ions through voltage-gated channels, a general leakage current I_L, as well as an external current (I) that is applied to the membrane by the experimenter.

Hodgkin and Huxley realised that the kinetics of the voltage-dependent ion channels are crucial to the rise and fall of the action potential. They formulated the membrane currents as,

$$
\begin{aligned}
I_{Na} &= g_{Na}\, m^3\, h\, (V - E_{Na}) \\
I_K &= g_K\, n^4\, (V - E_K) \\
I_L &= g_L\, (V - E_L)
\end{aligned}
\tag{6.1}
$$

where g_{Na}, g_K and g_L are the conductances of the ion channels and E_{Na}, E_K and E_L are the corresponding reversal potentials. The activation and inactivation of the voltage-gated sodium and potassium channels are represented by the gating variables m, h and n which all operate between 0 and 1. The kinetics of those gates are themselves defined by differential equations. Namely,

$$\tau_m(V) \frac{dm}{dt} = m_\infty(V) - m$$

$$\tau_h(V) \frac{dh}{dt} = h_\infty(V) - h$$

$$\tau_n(V) \frac{dn}{dt} = n_\infty(V) - n$$

where $m_\infty(V)$, $h_\infty(V)$ and $n_\infty(V)$ are the proportion of ion channels in the membrane that are activated (or inactivated) when the membrane potential is clamped at a given voltage. The timing characteristics of the gates are given by $\tau_m(V)$, $\tau_n(V)$ and $\tau_h(V)$ which are themselves voltage-dependent. Hodgkin and Huxley approximated those voltage relationships as,

$$m_\infty(V) = a_m/(a_m + b_m)$$
$$n_\infty(V) = a_n/(a_n + b_n) \qquad (6.2)$$
$$h_\infty(V) = a_h/(a_h + b_h)$$

and

$$\tau_m(V) = 1/(a_m + b_m)$$
$$\tau_n(V) = 1/(a_n + b_n) \qquad (6.3)$$
$$\tau_h(V) = 1/(a_h + b_h)$$

where a and b were established by empirical observation [30]. The toolbox version of the Hodgkin-Huxley model uses the contemporary expressions,

$$a_m = 0.1 * (V + 40)/(1 - \exp((-V - 40)/10))$$
$$a_n = 0.01 * (V + 55)/(1 - \exp((-V - 55)/10))$$
$$a_h = 0.07 * \exp((-V - 65)/20)$$
$$b_m = 4 * \exp((-V - 65)/18)$$
$$b_n = 0.125 * \exp((-V - 65)/80)$$
$$b_h = 1/(1 + \exp((-V - 35)/10))$$

described by Hansel, Mato and Meunier [22].

Visualising the Membrane Currents

The MATLAB ODE solvers are designed to return the time course of the state variables, in this case $V(t)$, $m(t)$, $h(t)$ and $n(t)$, but not any hidden variables that were used to compute the right-hand side of the differential equation. This creates a practical problem when we wish to inspect expressions such as the ionic currents (6.1) and the steady-state voltage relations (6.2). Whilst

in some cases it is possible to rearrange the differential equations to recast
those expressions as dynamic variables. It is often simpler to use auxiliary
plot functions to reconstruct the expressions post hoc. There are two basic
approaches, depending on whether or not the expressions rely on the computed solution.

Plotting the Steady-State Voltage Relations

The `VoltageGates` auxiliary function (Listing 6.3) plots $m_\infty(V)$, $h_\infty(V)$
and $n_\infty(V)$ for a fixed range of V (Figure 6.3). This particular auxiliary function does not utilise the solution data (`sol`) nor any of the model's parameters (`C, I, gNa, gK, gL, ENa, EK, EL`). It simply recapitulates the expressions for the steady-state voltage relations (6.2) in the auxiliary function itself (lines 7–16). This approach is appropriate when the plotted relations are
independent of the dynamic variables. In more complicated cases, it is pru-
dent to redefine the expressions as shared utility functions. This particular
function also returns the computed values in the `UserData` structure (lines
28–32) which makes them accessible to the workspace.

```
>> gui.panels.bdAuxiliary.UserData
   struct with fields:
        V: [1x201 double]
     minf: [1x201 double]
     hinf: [1x201 double]
     ninf: [1x201 double]
```

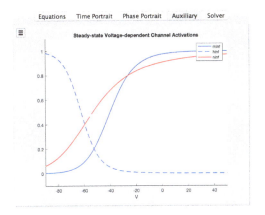

Fig. 6.3 The `Voltage-`
`Gates` auxiliary plot from
the *HodgkinHuxley* model.
It plots $m_\infty(V)$, $h_\infty(V)$ and
$n_\infty(V)$ for $-90 < V < 50$.

Listing 6.3 The `VoltageGates` auxiliary function from the *HodgkinHuxley* model. It plots the steady-state voltage-relations $m_\infty(V)$, $h_\infty(V)$ and $n_\infty(V)$ for a fixed range of V.

```
 1   function UserData = VoltageGates(ax,tt,sol,C,I,gNa,gK, ...
 2       gL,ENa,EK,EL)
 3
 4       % Voltage domain of interest
 5       V = linspace(-90,50,201);
 6
 7       % Steady-state Hodgkin-Huxley channel activations
 8       am = 0.1*(V+40)./(1-exp((-V-40)./10));
 9       an = 0.01*(V+55)./(1-exp((-V-55)./10));
10       ah = 0.07*exp((-V-65)./20);
11       bm = 4*exp((-V-65)./18);
12       bn = 0.125*exp((-V-65)./80);
13       bh = 1./(1+exp((-V-35)./10));
14       minf = am ./ (am + bm);
15       ninf = an ./ (an + bn);
16       hinf = ah ./ (ah + bh);
17
18       % Plot minf, ninf and hinf
19       plot(ax, V, minf , 'b-',   'Linewidth',1.5);
20       plot(ax, V, hinf , 'b--',  'Linewidth',1.5);
21       plot(ax, V, ninf , 'r-',   'Linewidth',1.5);
22       ylim(ax, [-0.1 1.1]);
23       xlim(ax, [-90 50]);
24       legend(ax,'minf','hinf','ninf');
25       title(ax,'Steady-state Voltage-dependent Activations');
26       xlabel(ax,'V');
27
28       % Make a copy of the data accessible to the workspace
29       UserData.V = V;
30       UserData.minf = minf;
31       UserData.hinf = hinf;
32       UserData.ninf = ninf;
33   end
```

Plotting the Ionic Currents

The `IonCurrents` auxiliary function (Listing 6.4) plots the membrane currents I_{Na}, I_K and I_L. The equations for these currents (6.1) depend on $V(t), m(t), h(t)$ and $n(t)$ which must be extracted from the solution structure (`sol`). The main loop (lines 11–24) iterates through the solution data and isolates the solution vector (`Y`) at each time point (`t`). The vector `Y` contains the four instantaneous values of V, m, h and n that are needed to reconstruct I_{Na}, I_K and I_L. Rather than duplicating equations (6.1) in the source code, the `IonCurrents` auxiliary function calls the `odefun` function (line 18)

Listing 6.4 The `IonCurrents` auxiliary function from the *HodgkinHuxley* model.

```
1  % Auxiliary function that plots the ion currents
2  function UserData = IonCurrents(ax,tt,sol,C,I,gNa,gK,gL, ...
3      ENa,EK,EL)
4
5      % Initialise UserData
6      UserData.t   = sol.x;
7      UserData.INa = NaN(size(sol.x));
8      UserData.IK  = NaN(size(sol.x));
9      UserData.IL  = NaN(size(sol.x));
10
11     % For each time step in the solution
12     for tindx = 1:numel(sol.x)
13         % get the solution data for the current time step
14         Y = sol.y(:,tindx);
15         t = sol.x(tindx);
16
17         % call the ODE to reconstruct the ionic currents
18         [~,INa,IK,IL] = odefun(t,Y,C,I,gNa,gK,gL,ENa,EK,EL);
19
20         % accumulate the results in UserData
21         UserData.INa(tindx) = INa;
22         UserData.IK(tindx)  = IK;
23         UserData.IL(tindx)  = IL;
24     end
25
26     % plot the ionic currents
27     plot(ax,UserData.t,UserData.INa,'r');
28     plot(ax,UserData.t,UserData.IK,'b');
29     plot(ax,UserData.t,UserData.IL,'g');
30     plot(ax,UserData.t,UserData.INa+UserData.IK+UserData.IL);
31     xlim(ax,UserData.t([1 end]));
32     ylim(ax,[-800 800]);
33     xlabel(ax,'time');
34     ylabel(ax,'current density');
35     legend(ax,'INa','IK','IL','combined');
36     title(ax,'Ionic Currents');
37  end
```

to reconstruct them exactly as they were computed by the solver. The extra outputs can easily be added to the `odefun`. Exposing `INa`, `IK` and `IL`,

```
function [dY,INa,IK,IL] = odefun(t,Y,...)
```

allows our auxiliary function to recapitulate the membrane currents exactly. This approach requires more programming effort but it guarantees that the auxiliary plot faithfully reconstructs the expressions that were used to compute the solution.

6.3 BOLD Haemodynamic Response Function

The Blood Oxygen Level Dependent (BOLD) Haemodynamic Response Function (HRF) (Figure 6.4) simulates the brain signal observed in functional Magnetic Resonance Imaging (fMRI). More specifically, it simulates the BOLD response to an impulse of neuronal activity within a region of the brain about the size of an fMRI voxel. That impulse response can later be applied to the activity signal of a neural mass model to obtain the corresponding fMRI signal. The *BOLDHRF* model uses an auxiliary plot function (Listing 6.5) to compute the impulse response (Figure 6.4) from the state variables $q(t)$ and $v(t)$. It also returns the computed response to the Auxiliary panel so that it is accessible to the workspace.

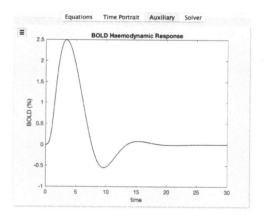

Fig. 6.4 The Auxiliary panel of the BOLDHRF model computes the haemodynamic response to an impulse of neural activity.

The Haemodynamic Model

The haemodynamic model combines the balloon model of the BOLD signal [6] with a dynamical model of the regional cerebral blood flow due to neuronal activity [19]. The balloon model defines the BOLD signal as

$$y(t) = V_0 \left(k_1 \left(1 - q(t) \right) + k_2 \left(1 - \frac{q(t)}{v(t)} \right) + k_3 \left(1 - v(t) \right) \right)$$

where $v(t)$ is the normalised volume of blood in the corresponding voxel of cerebral cortex and $q(t)$ is its normalised deoxyhaemoglobin content. Both variables are normalised with respect to their values at rest and so are dimensionless. Parameter $V_0 \approx 0.02$ is the resting blood volume fraction. Parameters $k_1 = 7E_0$, $k_2 = 2$ and $k_3 = (2E_0 - 0.2)$ weight the contributions of extra-

Listing 6.5 The BOLD auxiliary function from the BOLDHRF model.

```
1   function UserData = BOLD(ax,t,sol,V0,E0,tau0,tau1,...
2                             alpha,kappa,gamma,Z,ton,toff)
3       % extract solution
4       v = sol.y(1,:);      % blood volume
5       q = sol.y(2,:);      % deoxyhaemoglobin
6
7       % compute the BOLD signal
8       k1 = 7*E0;
9       k2 = 2;
10      k3 = 2*E0 - 0.2;
11      y = V0*(k1*(1-q) + k2*(1-q./v) + k3*(1-v));
12      t = sol.x;
13
14      % plot the BOLD signal
15      plot(ax, t, 100*y, 'color','k', 'LineWidth',1);
16      ylabel(ax, 'BOLD (%)');
17      xlabel(ax, 'time');
18      title(ax, 'BOLD Haemodynamic Response');
19
20      % make the data accessible to the workspace
21      UserData.t = t;
22      UserData.y = y;
23  end
```

and intravascular signals where $E_0 \approx 0.34$ is the resting oxygen extraction fraction [6].

The changes in blood volume and deoxyhaemoglobin are governed by the inflow (f_{in}) and outflow (f_{out}) of blood according to the coupled equations,

$$\tau_0 \, \dot{v} = f_{in} - f_{out}$$
$$\tau_0 \, \dot{q} = f_{in} \frac{E(f_{in})}{E_0} - f_{out} \frac{q}{v}$$

where $\tau_0 \approx 0.98$ seconds [19] is the mean transit time of blood through the compartment [6]. The extraction of oxygen from the blood during that transit is approximated by the function $E(f) = 1 - (1-E_0)^{1/f}$.

The outflow of blood is based on the windkessel model,

$$f_{out}(v) = v^{1/\alpha},$$

in which the outflow increases when the venous balloon is distended by higher blood volume. The stiffness of the balloon is estimated to be $\alpha \approx 0.38$

under steady-state conditions [21] and $\alpha \approx 0.33$ under normal fluctating conditions [19].

The inflow of blood is considered to be a linear function of the synaptic activity in the voxel [19]. It is characterised by the coupled linear equations,

$$\tau_1 \, \dot{f}_{in} = s$$
$$\tau_1 \, \dot{s} = u(t) - \kappa s - \gamma(f_{in} - 1)$$

where $s(t)$ is an unspecified vasodilatory signal that changes in response to both blood flow and neuronal activity $u(t)$. The decay rate of the of the signal is $\kappa \approx 1/1.54$ seconds and the time constant of the autoregulatory feedback is $\gamma \approx 1/2.46$ seconds [19]. The time constant is $\tau_1 = 1$ seconds.

Chapter 7
Command-line Tools

The command-line tools allow models to be solved in automated scripts without evoking the graphical user interface (bdGUI). This permits large parameter surveys to be conducted in batch mode and/or models to be embedded in third-party tools. The three main functions for this purpose are:

```
function sol = bdSolve(sys)
function [sys,sol] = bdEvolve(sys)
function [y,yp] = bdEval(sol,xint)
```

where bdSolve solves the initial value problem in the given sys structure and returns the solution as a sol structure. The bdEvolve function is similar except that it also returns a new sys structure where the initial values are replaced by the final values of the previous run. In both cases the bdEval function interpolates the solver's output at a given set of time points (xint) and returns the solution time series in y and the corresponding time-derivatives in yp. It is equivalent to MATLAB's deval function except that it also works for sol structures generated by third-party solvers (e.g. odeEul, sdeEM, sdeSH).

The exact format of the sol structure depends upon the solver routine that generated it. Nonetheless, all solvers return the following fields

```
sol =
struct with fields:
    solver: 'ode45'
         x: [1x126 double]
         y: [2x126 double]
     stats: [1x1 struct]
```

where sol.x contains the solver time steps and sol.y contains the corresponding solution values at each time step. It is acceptable to read these values directly from the sol structure without using bdEval but the time steps are not necessarily equidistant. The solver string identifies the solver routine that generated the `sol` structure. The `stats` structure contains various statistics about solver performance, such as the number of steps taken and the number of function evaluations.

```
sol.stats =
struct with fields:
    nsteps: 125
   nfailed: 0
   nfevals: 751
```

Other useful toolbox functions include `bdSetPar` and `bdGetPar` which manipulate the system parameters in a `sys` structure. The use of these functions is demonstrated in the next section. The complete syntax for all functions is given in Section 7.4.

7.1 Automating a model

Automating a model with the command-line tools is fairly straight-forward. We demonstrate the method with the `LinearODE` model (Listing 7.1). It has two state variables (x,y) and four scalar parameters (a,b,c,d). The procedure begins (line 2) by calling the `LinearODE` function to construct the `sys` structure for the model. The `bdSetPar` function (lines 5–8) is then used to set the parameters of the model $(a=1,b=-1,c=10,d=-2)$. These values are stored within the `pardef` field of the system structure. Likewise, the `bdSetVar` function (lines 11–12) is used to set the initial values of the ODE variables in the `vardef` field of the system structure. In this case, the x and y variables are both initialised with random values drawn from a uniform distribution. The `bdSolve` function (line 16) uses the `ode45` solver to integrate the model over the time span $0 \le t \le 10$. It returns the solution in the `sol` structure. The `bdEval` function (line 20) interpolates the solution at the time points specified in `tplot`. The script ends (lines 23-25) by plotting the computed solution (shown in Figure 7.1).

We remark that not all parameters and variables need to be explicitly initialised with `bdSetPar` and `bdSetVar`. Any that are omitted will retain their default values defined by `LinearODE`. Use the `bdGetPar` and `bdGetVar` functions to inspect the values in a system structure.

The approach illustrated in Listing 7.1 can be generalised to other types of differential equations. For DDEs the time lag parameters can be set using the `bdSetLag` function. For SDEs the `InitialStep` and `NoiseSources` options can be assigned directly to the `sys.sdeoption` data structure.

Listing 7.1 Running the `LinearODE` model using the command line tools.

```
1   % Construct the sys struct for the LinearODE model.
2   sys = LinearODE();
3
4   % set ODE parameters a=1, b=-1, c=10, d=-2
5   sys = bdSetPar(sys,'a',1);
6   sys = bdSetPar(sys,'b',-1);
7   sys = bdSetPar(sys,'c',10);
8   sys = bdSetPar(sys,'d',-2);
9
10  % set ODE variables x and y to random initial conditions
11  sys = bdSetVar(sys,'x',rand);
12  sys = bdSetVar(sys,'y',rand);
13
14  % integrate from t=0 to t=10 using the ode45 solver
15  tspan = [0 10];
16  sol = bdSolve(sys,tspan,@ode45);
17
18  % interpolate the results
19  tplot = 0:0.1:10;          % time domain for plotting
20  Y = bdEval(sol,tplot);     % interpolate the solution
21
22  % plot the result
23  plot(tplot,Y);
24  xlabel('time');
25  ylabel('x,y');
```

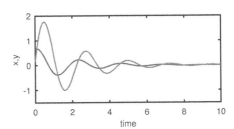

Fig. 7.1 The output produced by Listing 7.1 (lines 22–25).

7.2 Validating a model

The `bdSysCheck` function performs extensive validation checks on the contents of a `sys` structure. It is a valuable utility for debugging the system structures of newly-developed models. It checks the format of the fields within the `sys` struct and their legal combinations. It can also run the model on the relevant solvers to ensure that they operate as expected. Users are strongly encouraged to make frequent use of `bdSysCheck` when developing their own models. It offers the best chance of detecting errors in the system structure before running the model in `bdGUI` or `bdSolve`. The following example validates the system structure for the `LinearODE` model.

```
>> sys = LinearODE();
>> bdSysCheck(sys,'run','on')

sys struct format is OK
Calling Y = sys.odefun(t,Y0,a,b,c,d) where
   t is size [1  1]
   Y0 is size [2  1]
   a is size [1  1]
   b is size [1  1]
   c is size [1  1]
   d is size [1  1]
   Returns Y as size [2 1]
   sys.odefun format is OK
Calling sol = ode45(sys.odefun,Y0,sys.odeoption,
a,b,c,d) returns
   sol.x as size [1 127]
   sol.y as size [2 127]
Calling sol = ode23(sys.odefun,Y0,sys.odeoption,
a,b,c,d) returns
   sol.x as size [1 609]
   sol.y as size [2 609]
Calling sol = odeEul(sys.odefun,Y0,sys.odeoption,
a,b,c,d) returns
   sol.x as size [1 101]
   sol.y as size [2 101]
ALL TESTS PASSED OK
```

7.3 Calling the solver directly

In essence, the `bdSolve` function calls the given solver (e.g. `ode45`) and returns the `sol` structure to the caller. It is possible (albeit unnecessary) to call the solver directly without using `bdSolve`. The technique can be used as a last resort for debugging knotty problems in the system of equations. The syntax differs slightly for ODEs, DDEs and SDEs.

ODE solvers

The syntax for calling an ODE solver is

```
sol = ode45(sys.odefun, sys.tspan, y0, ...
                sys.odeoption, par{:});
```

where the initial conditions y0 are a monolithic column vector that can be obtained from the sys structure using the bdGetValues function.

```
y0 = bdGetValues(sys.vardef);
```

The model-specific parameters par are passed to ode45 as a cell array. The use of the {:} operator makes the contents of the cell array appear to the function as separate input parameters. The parameters themselves can be obtained directly from the sys.pardef array in a single operation.

```
par = {sys.pardef.value};
```

The enclosing braces are important because they cast the comma-separated list of values into a cell array.

DDE solvers

The syntax for calling a DDE solver is similar,

```
sol = dde23(sys.ddefun, lag, y0, sys.tspan, ...
                sys.ddeoption, par{:});
```

where the lag parameters and initial conditions y0 are both monolithic column vectors.

```
lag = bdGetValues(sys.lagdef);
y0  = bdGetValues(sys.vardef);
```

The model-specific parameters par are obtained from the sys structure using the same technique described above.

```
par = {sys.pardef.value};
```

SDE solvers

Likewise, the syntax for calling a SDE solver is

```
sol = sdeEM(sys.sdeF, sys.sdeG, sys.tspan, ...
                y0, sys.sdeoption, par{:});
```

where the initial conditions y0 and the model-specific parameters par are all obtained using the same methods described above. The major difference being that SDE solvers require two function handles (@sdeF and @sdeG) instead of one.

7.4 Syntax Reference

bdEval

```
[y,yp] = bdEval(sol,xint)
[y,yp] = bdEval(sol,xint,idx)
[y,yp] = bdEval(sol,xint,sys,'name')
```

Evaluates the solution (sol) previously returned by a solver routine by interpolating it at the given time points (xint). This function is equivalent to the MATLAB deval function except that it also works for sol structures generated by third-party solvers. Like deval, it returns the values of the solution (y) and its first derivative (yp) evaluated at the time points given in xint. The optional idx parameter specifies which components (rows) of the solution to return. Alternatively, the corresponding state variable in the system structure (sys) can be specified by name, whereupon the values returned in y and yp are automatically reshaped to the appropriate dimensions.

bdEvolve

```
[sys,sol] = bdEvolve(sys)
[sys,sol] = bdEvolve(sys,rep)
[sys,sol] = bdEvolve(sys,rep,tspan)
[sys,sol] = bdEvolve(sys,rep,tspan,@solverfun)
[sys,sol] = bdEvolve(sys,rep,tspan,@solverfun,...
                                        'solvertype')
```

Solves the model without invoking the graphical user interface in the same way as bdSolve. In this case it also returns an updated copy of the sys structure in which the initial conditions (sys.vardef.value) have been replaced with the final states of the last simulation run. The purpose of the function is to evolve a solution over multiple simulation runs where the rep parameter specifies the number of runs. It defaults to rep=1 when omitted. All other parameters are the same as for bdSolve. The returned solution structure (sol) is that of the final run.

bdGetLag

```
[val,idx] = bdGetLag(sys,'name')
```

Gets the value of the named lag parameter in the system structure (sys). The value of the lag parameter is returned in val. Its position in the sys.lagdef array is returned in idx. Both val and idx are returned empty if no matching name was found. If the name parameter is omitted then bdGetLag(sys) will list the names of all lag parameters in sys.

bdGetPar

```
[val,idx] = bdGetPar(sys,'name')
```

Gets the value of the named parameter in the system structure (`sys`). The value of the parameter is returned in `val`. Its position in the `sys.pardef` array is returned in `idx`. Both `val` and `idx` are returned empty if no matching name was found. If the `name` parameter is omitted then `bdGetPar(sys)` will list the names of all parameters in `sys`.

bdGetValue

```
[val,idx] = bdGetValue(xxxdef,'name')
```

Gets a named value from `xxxdef` which is either a struct array of system parameters (`sys.pardef`) or system variables (`sys.vardef`) or lag parameters (`sys.lagdef`). The `name` string identifies the target entry in the struct array. The value of the entry and its array index are returned in `val` and `idx` respectively. Both are returned empty if no matching name was found.

bdGetValues

```
Y = bdGetValues(xxxdef)
```

Returns the values of all system parameters as one monolithic column vector where `xxxdef` is a struct array of system parameters (`sys.pardef`) or system variables (`sys.vardef`) or lag parameters (`sys.lagdef`).

bdGetVar

```
[val,idx,solidx] = bdGetVar(sys,'name')
```

Gets the value of the named state variable in the system structure (`sys`). The value of the variable is returned in `val`. Its position in the `sys.vardef` array is returned in `idx`. Its corresponding row position in the solver output (`sol.y`) is returned in `solidx`. All output parameters are returned empty if no matching name was found. If the `name` parameter is omitted then `bdGetVar(sys)` will list the names of all state variables in `sys`.

bdGUI

```
gui = bdGUI()
gui = bdGUI(sys)
gui = bdGUI(sys,'sol',sol)
```

The graphical user interface for the Brain Dynamics Toolbox. It loads and runs the model defined by the given system structure (`sys`). If no `sys` parameter is given then `bdGUI` prompts the user to load a *mat* file that contains a `sys` structure. If that *mat* file also contains a `sol` structure then it too is loaded. An existing `sol` structure can also be loaded directly as an input parameter. In all cases, ensure that the MATLAB search path includes the functions that are used in the `sys` structure function handles (e.g. `sys.odefun`) otherwise the model will fail to load. The `gui` object that is returned by `bdGUI` can be used to control the graphical user interface from the workspace (See Section 1.5).

bdJacobian

```
J = bdJacobian(sys,t,Y)
```

Returns the Jacobian matrix $[\partial F_i/\partial Y_j]$ of the dynamical system evaluated at the point `Y` at time `t`. The Jacobian is only defined for ODEs. It is computed using the model's `sys.odeoption.Jacobian` function if it exists. Otherwise it is estimated numerically using finite differences. The `t` parameter can safely be defined as `NaN` for autonomous (time independent) dynamical systems.

bdSetLag

```
sys = bdSetLag(sys,'name',val)
```

Sets the value of the named lag parameter and returns the updated system structure (`sys`). The `name` string identifies the target entry in the `sys.lagdef` array.

bdSetPar

```
sys = bdSetPar(sys,'name',val)
```

Sets the value of the named system parameter and returns the updated system structure (`sys`). The `name` string identifies the target entry in the `sys.pardef` array.

bdSetValue

```
yyydef = bdSetValue(xxxdef,'name',val)
```

Sets the value of a named entry in `xxxdef` which is either a struct array of system parameters (`sys.pardef`) or system variables (`sys.vardef`) or lag parameters (`sys.lagdef`). The `name` string identifies the target entry to be updated and `val` must be a numeric scalar, vector or matrix. The updated struct array is returned in `yyydef`.

bdSetValues

```
yyydef = bdSetValues(xxxdef,Y)
```

Sets the values of all system parameters or variables from a monolithic column vector. The `xxxdef` parameter is a struct array of either system parameters (`sys.pardef`) or system variables (`sys.vardef`) or lag parameters (`sys.lagdef`). The vector `Y` must have one element for each value in the `xxxdef` array of structs. The operation is the inverse of `Y=bdGetValues(xxxdef)`.

bdSetVar

```
sys = bdSetVar(sys,'name',val)
```

Sets the value of the named system variable and returns the updated system structure (`sys`). The `name` string identifies the target entry in the `sys.vardef` array.

bdSolve

```
sol = bdSolve(sys)
sol = bdSolve(sys,tspan)
sol = bdSolve(sys,tspan,@solverfun)
sol = bdSolve(sys,tspan,@solverfun,'solvertype')
```

Solves the model defined by the given `sys` structure and returns the solver output in `sol`. The graphical user interface is not invoked. The `tspan` argument defines the time span of the integration [`t0 t1`]. It defaults to `sys.tspan` if omitted. The `solverfun` argument is a function handle to the relevant solver routine (e.g. `@ode45`). If omitted, it defaults to the first solver identified in the `sys` structure (`odefun`, `ddefun` or `sdefun`). The `solvertype` string identifies the type of the solver routine (`'odesolver'`, `'ddesolver'` or `'sdesolver'`). It is only needed when `solverfun` refers to a solver routine that is unknown to the toolbox, such as a user-defined solver. The `sol` structure is the same as that returned by the MATLAB solvers (e.g. `ode45`, `dde23`). Use the `bdEval` function to obtain a time series from it.

bdSysCheck

```
bdSysCheck(sys)
bdSysCheck(sys,'run','on')
```

Checks the validity of a model's system structure. This debugging utility runs an extensive suite of tests on the format of the fields in the given `sys` structure. It detects missing fields and illegal field combinations and prints warnings in the MATLAB command window. If the `run` option is `'on'` then it also runs the model with the relevant solvers — using the default values defined in the `sys` structure. We recommend using `bdSysCheck` routinely during the development of any new model. See Section 7.2.

References

1. Aburn M.J. (2017) Critical fluctuations and coupling of stochastic neural mass models. Ph.D. thesis, University of Queensland.
2. Bower J.M., Beeman D. (1998) The book of Genesis: exploring realistic neural models with the General Neural Simulation System. Telos, Springer, New York.
3. Breakspear M., Heitmann S., Daffertshofer A. (2010) Generative models of cortical oscillations: Neurobiological implications of the Kuramoto model. Frontiers in Human Neuroscience 4, 190.
4. Breakspear M., Terry J. (2002) Detection and description of non-linear interdependence in normal multichannel human EEG data. Clinical Neurophysiology 113(5), 753.
5. Breakspear M., Terry J.R., Friston K.J. (2003) Modulation of excitatory synaptic coupling facilitates synchronization and complex dynamics in a biophysical model of neuronal dynamics. Network (Bristol, England) 14(4), 703–32. ISSN 0954898.
6. Buxton R.B., Wong E.C., Frank L.R. (1998) Dynamics of blood flow and oxygenation changes during brain activation: The balloon model. Magnetic Resonance in Medicine 39(6), 855–864. ISSN 07403194, 15222594. doi:10.1002/mrm.1910390602.
7. Carnevale N.T., Hines M.L. (2006) The NEURON book. Cambridge University Press.
8. Clewley R. (2012) Hybrid Models and Biological Model Reduction with PyDSTool. PLOS Computational Biology 8(8), e1002628. ISSN 1553-7358. doi:10.1371/journal.pcbi.1002628.
9. Dafilis M.P., Frascoli F., Cadusch P.J., Liley D.T. (2009) Chaos and generalised multistability in a mesoscopic model of the electroencephalogram. Physica D: Nonlinear Phenomena 238(13), 1056–1060. ISSN 01672789. doi:10.1016/j.physd.2009.03.003.
10. Dankowicz H., Schilder F. (2013) Recipes for Continuation. SIAM. ISBN 978-1-61197-256-6.
11. Dhooge A., Govaerts W., Kuznetsov Y.A. (2003) MATCONT: a MATLAB package for numerical bifurcation analysis of ODEs. ACM Transactions on Mathematical Software (TOMS) 29(2), 141–164.
12. Doedel E.J., Champneys A.R., Fairgrieve T.F., Kuznetsov Y.A., Sandstede B., Wang X. (1998). AUTO 97: Continuation and bifurcation software for ordinary differential equations (with HomCont).
13. Ermentrout B. (2002) Simulating, analyzing, and animating dynamical systems: a guide to XPPAUT for researchers and students. Society for Industrial Mathematics.
14. Ermentrout G.B., Terman D.H. (2010) Mathematical foundations of neuroscience, volume 35. Springer.

15. Fisher R.A. (1937) The Wave of Advance of Advantageous Genes. Annals of Eugenics 7(4), 355–369. ISSN 2050-1439. doi:10.1111/j.1469-1809.1937.tb02153.x.

16. FitzHugh R. (1955) Mathematical models of threshold phenomena in the nerve membrane. Bulletin of Mathematical Biology 17(4), 257–278.

17. Freyer F., Roberts J.A., Ritter P., Breakspear M. (2012) A Canonical Model of Multistability and Scale-Invariance in Biological Systems. PLoS Computational Biology 8(8), e1002634. ISSN 1553-7358. doi:10.1371/journal.pcbi.1002634.

18. Gardiner C. (2009) Stochastic Methods: A Handbook for the Natural and Social Sciences. Springer, 4th edition. ISBN 978-3-540-70712-7.

19. Glaser D.E., Friston K.J., Mechelli A., Turner R., Price C.J. (2003) Haemodynamic modelling. In Human Brain Function, volume 2, pages 823–842. Springer Science & Business Media.

20. Goodman D.F.M., Brette R. (2013) BRIAN simulator. Scholarpedia 8(1), 10883. ISSN 1941-6016. doi:10.4249/scholarpedia.10883.

21. Grubb R.L., Raichle M.E., Eichling J.O., Ter-Pogossian M.M. (1974) The effects of changes in PaCO2 cerebral blood volume, blood flow, and vascular mean transit time. Stroke 5(5), 630–639.

22. Hansel D., Mato G., Meunier C. (1993) Phase Dynamics for Weakly Coupled Hodgkin-Huxley Neurons. Europhysics Letters (EPL) 23(5), 367–372. ISSN 0295-5075. doi:10.1209/0295-5075/23/5/011.

23. Heitmann S., Aburn M.J., Breakspear M. (2017) The Brain Dynamics Toolbox for Matlab. bioRxiv page 219329. doi:10.1101/219329.

24. Heitmann S., Breakspear M. (2018) Putting the dynamic back into dynamic functional connectivity. Network Neuroscience pages 1–61. ISSN 2472-1751. doi:10.1162/NETN_a_00041.

25. Heitmann S., Ermentrout G.B. (2015) Synchrony, waves and ripple in spatially coupled Kuramoto oscillators with Mexican hat connectivity. Biological Cybernetics 109(3), 333–347. ISSN 0340-1200, 1432-0770. doi:10.1007/s00422-015-0646-6.

26. Heitmann S., Ermentrout G.B. (2020) Direction-selective motion discrimination by traveling waves in visual cortex. PLOS Computational Biology 16(9), e1008164. ISSN 1553-7358. doi:10.1371/journal.pcbi.1008164.

27. Heitmann S., Rule M., Truccolo W., Ermentrout B. (2017) Optogenetic Stimulation Shifts the Excitability of Cerebral Cortex from Type I to Type II: Oscillation Onset and Wave Propagation. PLOS Computational Biology 13(1), e1005349. ISSN 1553-7358. doi:10.1371/journal.pcbi.1005349.

28. Hindmarsh J.L., Rose R.M. (1984) A model of neuronal bursting using three coupled first order differential equations. Proceedings of the Royal Society of London B: Biological Sciences 221(1222), 87–102.

29. Hlinka J., Coombes S. (2012) Using computational models to relate structural and functional brain connectivity: Relating structural and functional brain connectivity. European Journal of Neuroscience 36(2), 2137–2145. ISSN 0953816X. doi:10.1111/j.1460-9568.2012.08081.x.

30. Hodgkin A.L., Huxley A.F. (1952) A quantitative description of membrane current and its application to conduction and excitation in nerve. The Journal of Physiology 117(4), 500–544. ISSN 0022-3751.

31. Hopfield J.J. (1982) Neural networks and physical systems with emergent collective computational abilities. Proceedings of the national academy of sciences 79(8), 2554–2558.

32. Jacobs K. (2010) Stochastic processes for physicists: understanding noisy systems. Cambridge University Press.

33. Jirsa V.K., Stacey W.C., Quilichini P.P., Ivanov A.I., Bernard C. (2014) On the nature of seizure dynamics. Brain 137(8), 2210–2230. ISSN 1460-2156, 0006-8950. doi:10.1093/brain/awu133.

34. Kloeden P.E., Platen E. (1992) Numerical solution of stochastic differential equations springer-verlag. Number 23 in Applications of Mathematics. Springer-Verlag. ISBN 3-540-54062-8.

35. Kotter R. (2004) Online retrieval, processing, and visualization of primate connectivity data from the CoCoMac database. Neuroinformatics 2(2), 127–144.

36. Kottwitz S. (2011) LaTeX beginner's guide. Packt Publishing Ltd.

37. Kuramoto Y. (1984) Chemical oscillations, waves, and turbulence. Dover Publications, Mineola, New York.

38. Lamport L. (1994) LaTeX: A Document Preparation System, 2nd Edition. Addison-Wesley Professional., 2nd edition. ISBN 978-0-201-52983-8.

39. Lecar H. (2007) Morris-Lecar model. Scholarpedia 2(10), 1333. ISSN 1941-6016. doi:10.4249/scholarpedia.1333.

40. Lorenz E.N. (1963) Deterministic nonperiodic flow. Journal of the atmospheric sciences 20(2), 130–141.

41. Marple L. (1999) Computing the discrete-time analytic signal via FFT. IEEE Transactions on Signal Processing 47(9), 2600–2603. ISSN 1053-587X. doi: 10.1109/78.782222.

42. Maruyama G. (1955) Continuous Markov processes and stochastic equations. Rendiconti del Circolo Matematico di Palermo 4(1), 48–90.

43. Morris C., Lecar H. (1981) Voltage oscillations in the barnacle giant muscle fiber. Biophysical Journal 35(1), 193–213. ISSN 0006-3495. doi:10.1016/S0006-3495(81) 84782-0.

44. Nagumo J., Arimoto S., Yoshizawa S. (1962) An Active Pulse Transmission Line Simulating Nerve Axon. Proceedings of the IRE 50(10), 2061–2070. ISSN 0096-8390. doi:10.1109/JRPROC.1962.288235.

45. Othmer H.G. (1997) Signal transduction and second messenger systems. In Case studies in mathematical modeling - ecology, physiology, and cell biology, page 101. Prentice Hall, Englewood Cliffs, New Jersey. ISBN 978-0-13-574039-2.

46. Othmer H.G., Tang Y. (1993) Oscillations and Waves in a Model of InsP3-Controlled Calcium Dynamics. In H.G. Othmer, P.K. Maini, J.D. Murray, editors, Experimental and Theoretical Advances in Biological Pattern Formation, pages 277–300. Springer US, Boston, MA. ISBN 978-1-4613-6033-9 978-1-4615-2433-5. doi: 10.1007/978-1-4615-2433-5_25.

47. Park Y., Heitmann S., Ermentrout B. (2017) The utility of phase models in studying neuronal synchronization. In Computational models of brain and behavior, pages 493–504. John Wiley & Sons, 1st edition. ISBN 978-1-119-15906-3. Ed Ahmed A Moustafa.

48. Pospischil M., Toledo-Rodriguez M., Monier C., Piwkowska Z., Bal T., Frgnac Y., Markram H., Destexhe A. (2008) Minimal Hodgkin-Huxley type models for different classes of cortical and thalamic neurons. Biological Cybernetics 99(4-5), 427–441. ISSN 0340-1200, 1432-0770. doi:10.1007/s00422-008-0263-8.

49. Rand R.H., Holmes P.J. (1980) Bifurcation of periodic motions in two weakly coupled van der Pol oscillators. International Journal of Non-Linear Mechanics 15(4-5), 387–399.

50. Roberts J.A., Friston K.J., Breakspear M. (2017) Clinical Applications of Stochastic Dynamic Models of the Brain, Part I: A Primer. Biological Psychiatry: Cognitive Neuroscience and Neuroimaging ISSN 24519022. doi:10.1016/j.bpsc.2017.01.010.

51. Ruemelin W. (1982) Numerical treatment of stochastic differential equations. SIAM Journal on Numerical Analysis 19(3), 604–613.

52. Sakaguchi H., Kuramoto Y. (1986) A soluble active rotater model showing phase transitions via mutual entertainment. Progress of Theoretical Physics 76(3), 576–581.

53. Sanz-Leon P., Knock S.A., Spiegler A., Jirsa V.K. (2015) Mathematical framework for large-scale brain network modeling in The Virtual Brain. Neuroimage 111, 385–430.

54. Sanz Leon P., Knock S.A., Woodman M.M., Domide L., Mersmann J., McIntosh A.R., Jirsa V. (2013) The Virtual Brain: a simulator of primate brain network dynamics. Frontiers in Neuroinformatics 7. ISSN 1662-5196. doi:10.3389/fninf.2013.00010.

55. Shampine L.F., Gladwell I., Thompson S. (2003) Solving ODEs with Matlab. Cambridge University Press.

56. Smythe J., Moss F., McClintock P.V., Clarkson D. (1983) Ito versus Stratonovich revisited. Physics Letters A 97(3), 95–98.

57. Strogatz S.H. (1994) Nonlinear dynamics and chaos. Westview Pr.

58. Swift J., Hohenberg P.C. (1977) Hydrodynamic fluctuations at the convective instability. Physical Review A 15(1), 319–328. doi:10.1103/PhysRevA.15.319.

59. The Mathworks (2017). MATLAB & Simulink: Text with mathematical expression using LaTeX.

60. Tsodyks M.V., Skaggs W.E., Sejnowski T.J., McNaughton B.L. (1997) Paradoxical Effects of External Modulation of Inhibitory Interneurons. Journal of Neuroscience 17(11), 4382–4388. ISSN 0270-6474, 1529-2401. doi:10.1523/JNEUROSCI.17-11-04382.1997.

61. Van der Pol B. (1934) The nonlinear theory of electric oscillations. Proceedings of the Institute of Radio Engineers 22(9), 1051–1086.

62. Van Kampen N.G. (1981) Ito versus Stratonovich. Journal of Statistical Physics 24(1), 175–187.

63. Van Kampen N.G. (1992) Stochastic processes in physics and chemistry, volume 1. Elsevier.

64. Wille D.R., Baker C.T. (1992) DELSOL - a numerical code for the solution of systems of delay-differential equations. Applied Numerical Mathematics 9(3-5), 223–234.

65. Wilson H.R., Cowan J.D. (1972) Excitatory and inhibitory interactions in localized populations of model neurons. Biophysical Journal 12(1), 1–24.

66. Wilson H.R., Cowan J.D. (1973) A mathematical theory of the functional dynamics of cortical and thalamic nervous tissue. Kybernetik 13(2), 55–80.

Index

www.ingramcontent.com/pod-product-compliance
Lightning Source LLC
Chambersburg PA
CBHW041634050326
40689CB00024B/4961